Springer Theses

Recognizing Outstanding Ph.D. Research

For further volumes:
http://www.springer.com/series/8790

Aims and Scope

The series "Springer Theses" brings together a selection of the very best Ph.D. theses from around the world and across the physical sciences. Nominated and endorsed by two recognized specialists, each published volume has been selected for its scientific excellence and the high impact of its contents for the pertinent field of research. For greater accessibility to non-specialists, the published versions include an extended introduction, as well as a foreword by the student's supervisor explaining the special relevance of the work for the field. As a whole, the series will provide a valuable resource both for newcomers to the research fields described, and for other scientists seeking detailed background information on special questions. Finally, it provides an accredited documentation of the valuable contributions made by today's younger generation of scientists.

Theses are accepted into the series by invited nomination only and must fulfill all of the following criteria

- They must be written in good English.
- The topic should fall within the confines of Chemistry, Physics and related interdisciplinary fields such as Materials, Nanoscience, Chemical Engineering, Complex Systems and Biophysics.
- The work reported in the thesis must represent a significant scientific advance.
- If the thesis includes previously published material, permission to reproduce this must be gained from the respective copyright holder.
- They must have been examined and passed during the 12 months prior to nomination.
- Each thesis should include a foreword by the supervisor outlining the significance of its content.
- The theses should have a clearly defined structure including an introduction accessible to scientists not expert in that particular field.

Kaspar Sakmann

Many-Body Schrödinger Dynamics of Bose–Einstein Condensates

Doctoral Thesis accepted by Institute
of Physical Chemistry, Theoretical Chemistry,
Heidelberg University, Germany

Author
Dr. Kaspar Sakmann
Theoretical Chemistry Group
Institute of Physical Chemistry
Heidelberg University
Im Neuenheimer Feld 229
69120 Heidelberg
Germany
e-mail:
kaspar.sakmann@pci.uni-heidelberg.de

Supervisor
Prof. Dr. Lorenz S. Cederbaum
Theoretical Chemistry Group
Institute of Physical Chemistry
Heidelberg University
Im Neuenheimer Feld 229
69120 Heidelberg
Germany
e-mail:
lorenz.cederbaum@pci.uni-heidelberg.de

ISSN 2190-5053
ISBN 978-3-642-27105-2
DOI 10.1007/978-3-642-22866-7
Springer Heidelberg Dordrecht London New York

e-ISSN 2190-5061
ISBN 978-3-642-22866-7 (eBook)

Softcover reprint of the hardcover 1st edition 2011

Cover design: eStudio Calamar, Berlin/Figueres

Printed on acid-free paper

Springer is part of Springer Science+Business Media (www.springer.com)

Supervisor's Foreword

The field of ultracold quantum gases has fascinated physicists ever since the predictions of Bose and Einstein, but the greatest boost in the field set in after the first experimental realizations of Bose–Einstein condensates in 1995. The great attention that Bose–Einstein condensates have received since then is due to the following facts. Bose–Einstein condensates are macroscopic in size, but still follow the laws of quantum mechanics. They constitute a new state of matter. Bose–Einstein condensates are prepared in nearly perfect isolation which makes it possible to measure observables, such as the density of the many-body wave function directly in experiments. Last but not least there is an exceptional degree of control over the parameters that determine these systems. For example, the interaction strength, the number of particles and the external trapping potential can be controlled almost at will.

Thereby, Bose–Einstein condensates offer a prime laboratory for the investigation of many-body quantum physics. Most studies of Bose–Einstein condensates focus on idealized descriptions, usually in terms of the Gross–Pitaevskii mean-field equation, often also referred to as the nonlinear Schrödinger equation, or the Bose–Hubbard many-body model. However, the real strength of experimental and theoretical Bose–Einstein condensate research is the possibility to see the full many-body Schrödinger equation at work.

For a long time it appeared to be impossible to solve the time-dependent many-body Schrödinger equation for Bose–Einstein condensates in experimentally relevant situations. With the development of the Multiconfigurational Time-Dependent Hartree for Bosons (MCTDHB) method such studies are becoming feasible. In this thesis the MCTDHB method was used to solve the many-body Schrödinger numerically. Most notable are the results for a bosonic Josephson junction, a system with a long history of theoretical and experimental interest. These are the first and so far the only exact results in literature about bosonic Josephson junctions with a mesoscopic number of particles. A comparison to Gross–Pitaevskii theory and the Bose–Hubbard model proved that both of these approximations fail to describe bosonic Josephson Junctions in a regime where they are commonly believed to be valid. This has far-reaching consequences since

Gross–Pitaevskii theory and the Bose–Hubbard model are often used and work-horses in the field, not only in this context.

MCTDHB stimulated further research about possibilities to improve the Bose–Hubbard model. The second exceptional result of this thesis is that all lattice models of condensed matter physics that employ Wannier functions can be improved in an optimal way, and it is shown here how this can be achieved. The so obtained optimal lattice models are superior to their conventional counterparts, as is shown by direct comparison of the optimal and the conventional Bose–Hubbard model to exact results of the Schrödinger equation in a quantum quench scenario.

Heidelberg, June 2011 Prof. Dr. Lorenz S. Cederbaum

Acknowledgments

Many people have supported me in recent years and I am very grateful to them. First of all, I would like to thank my supervisor Prof. Lorenz Cederbaum. He was always able to separate the important from the unimportant. His guidance and foresight have helped me in staying focused on the true many-body physics of Bose–Einstein condensates. The friendly manner in which he is leading our group makes for a great working atmosphere. I have shared many memorable moments with Dr. Alexej Streltsov and Prof. Ofir Alon. I have learned a great deal from them. By now we have spent many years together and our scientific discussions were always very inspiring and helpful. Last but not least, I would like to thank my parents Christiane and Bert Sakmann and my lovely wife Lea-Ann.

Heidelberg, June 2011 Dr. Kaspar Sakmann

Supplementary Material

This thesis relies heavily on the MCTDHB method to solve the time-dependent many-body Schrödinger equation for bosons. People interested in MCTDHB can learn more about it at http://MCTDHB.uni-hd.de. In the framework of this thesis an Open-Source MCTDHB package was developed. Any researcher interested in trying out MCTDHB is welcome to download the OpenMCTDHB package from http://OpenMCTDHB.uni-hd.de.

Contents

Chapter 1
Introduction

In this chapter we will briefly review the path that lead to Bose–Einstein condensation in ultracold gases without going too much into the details of the language and the concepts that are nowadays commonly used in the field. No attempt is made here to be rigorous. For definitions and explanations we refer the reader to the references given and Chaps. 2–4.

1.1 The Path to Bose–Einstein Condensation

Bose–Einstein condensation was first proposed as a theoretical concept in the first half of the last century. When the Indian physicist Satayendra N. Bose investigated the statistics of photons, he discovered that the thermal distribution of photons is not of the Maxwell–Boltzmann type [1]. Nowadays, particles that obey the distribution function derived by Bose are known as bosons. Albert Einstein was impressed by Bose's work and extended it to a gas of massive, noninteracting particles [2, 3]. Einstein realized that for sufficiently low temperatures a large fraction of particles would occupy the state of lowest energy. At absolute zero temperature all particles would "condense" into the lowest energy state and hence behave all in the same manner. The idea of Bose–Einstein condensation was born and the search for Bose–Einstein condensates began.

The first candidate for a physical system exhibiting Bose–Einstein condensation was superfluid liquid ^{4}He, as suggested by London in 1938 [4, 5]. However, interactions between particles in superfluid liquid ^{4}He are strong, in stark contrast to the noninteracting particles in Einstein's model. Already back then it was expected that interactions would strongly modify the physics of Bose–Einstein condensation. This is indeed the case. Theoretical results and modern experiments suggest that the fraction of condensed particles in superfluid liquid ^{4}He is no larger than about 7%, even at absolute zero temperature [6–9].

The search for Bose–Einstein condensates therefore soon extended to other systems, see Ref. [10] for an overview. In the 1980s novel laser and magnetic based

K. Sakmann, *Many-Body Schrödinger Dynamics of Bose–Einstein Condensates*,
Springer Theses, DOI: 10.1007/978-3-642-22866-7_1,

cooling techniques were developed that allowed experimentalists to cool dilute gases of neutral atoms down to extremely low temperatures, see Refs. [11–13] for overviews of these techniques. After a long spell of nearly successful experiments Carl Wieman and collaborators envisaged a path towards Bose–Einstein condensation in dilute, atomic alkali gases [14, 15]. This path eventually proved to be successful. Two main ingredients of this procedure were the creation of an extremely good vacuum and evaporative cooling. In 1995 Bose–Einstein condensation in dilute alkali gases was achieved for the first time in a series of experiments using Rubidium in the group of Eric Cornell and Carl Wieman, Lithium in the group of Randall Hulet and Sodium in the group of Ketterle [16–18]. However, according to the Nobel committee [19] unambiguous results of Randall Hulet's group were published only in 1997 [20]. Only 6 years after the first realization of Bose–Einstein condensates in experiments, Wieman, Cornell and Ketterle were awarded the Nobel Prize in Physics "for the achievement of Bose–Einstein condensation in dilute gases of alkali atoms, and for early fundamental studies of the properties of the condensates" [21]. For a more detailed account of the history of the first Bose–Einstein condensates, see Refs. [19, 22].

In the years after 1995 two major new developments accelerated progress in the field of ultracold atoms. The first was the possibility to tune the interaction strength between particles by using Feshbach resonances [23, 24]. Thus, it became possible to go from weakly to strongly interacting Bose–Einstein condensates. The second development was progress in the shaping of trap geometries. While many of the first experiments were carried out in single-well traps, sophisticated trapping techniques allowed to shape almost arbitrary traps. In particular, traps with several potential minima soon became popular. Nowadays, multi-well traps and whole lattices of wells are commonly used in experiments [25–28]. Through the combination of the above developments, it even became possible to strongly confine the atoms in some of the spatial dimensions and thereby to realize quasi two- and one-dimensional Bose–Einstein condensates. The range of physical phenomena accessible to ultracold gases is vast, for example the interference of matter waves, tunneling, Josephson-like effects and strongly correlated bosons to name just a few. The field has also extended to ultracold fermions and the formation of ultracold molecules. Recent overviews of the developments in the field are given in Refs. [25–30].

1.2 Theories for Bose–Einstein Condensates

At the beginning of the theory of Bose–Einstein condensates is the ideal Bose gas. For the reader not familiar with Bose–Einstein condensation at all, it is recommendable to begin with the ideal Bose gas which is treated in standard textbooks on statistical mechanics, such as Ref. [31]. Alternatively, Ref. [32] contains a good chapter on the ideal Bose-gas. In the framework of the ideal Bose-gas the main problem is one of statistical mechanics: once the single particle problem is solved, questions such as the transition temperature and thermodynamical distribution functions can be computed. Although the ideal gas model is at the origin of the theory of Bose–Einstein condensation, it is too crude to describe any of the experiments to date.

One of the most important features of ultracold trapped Bose gases is that they are highly inhomogeneous, interacting many-body systems of finite size. The size and shape of the condensates are generally determined by the trap geometry, the number of particles and the interparticle interaction. A variation of any of these parameters can have large effects on the properties of the condensates.

The role of interparticle interactions is particularly important. An early experiment using about 80,000 sodium atoms showed that the density distribution the condensate was at least three to four times broader than that of a gas of noninteracting particles [33]. Interactions in ultracold Bose gases are therefore not negligible and have to be taken into account in theoretical treatments. Interactions lead to new collective effects which are very exciting, but also greatly complicate the theory of Bose–Einstein condensates. In fact, the literature on Bose–Einstein condensates consists for the largest part of works that address the behaviour of interacting Bose–Einstein condensates at zero temperature, i.e. the problem solved is the quantum mechanical many-boson problem and not one of statistical mechanics.

The equation that governs all properties of Bose–Einstein condensates of dilute, atomic gases is the interacting many-body Schrödinger equation, which will be discussed in Chap. 2. We assume that the reader is familiar with the formalism of second quantization to the extent that it is treated, e.g. in the first three chapters of Ref. [34]. The many-body Schrödinger equation is very difficult to solve, even for few particles, and approximations are usually indispensable.

Certainly the most popular of these approximative methods is the celebrated Gross–Pitaevskii theory which was developed independently by Gross and Pitaevskii in 1961 [35–37]. We will derive the Gross–Pitaevskii equation in Chap. 3 as a special case of a more general method to solve the many-body Schrödinger equation. Using Gross–Pitaevskii theory it is possible to investigate inhomogeneous, interacting condensates in arbitrary trap geometries. At first sight it seems to be a formidable general purpose theory for Bose–Einstein condensates. The only assumption on which Gross–Pitaevskii theory rests, is that the quantum state is fully condensed at all times. If the many-body Schrödinger dynamics of a Bose–Einstein condensate remains condensed at all times, there is no difference between the solution of the many-body Schrödinger equation and that of the Gross–Pitaevskii equation. However, the assumption of a fully condensed state is also the major drawback of Gross–Pitaevskii theory, since it is impossible to tell beforehand for a given system whether the assumption is justified or not, without going beyond Gross–Pitaevskii theory. The previously mentioned example of superfluid liquid ^{4}He, where no more than about 7% of all particles are believed to be condensed, proves that this question is a very relevant one in interacting Bose systems.

Another important aspect of trapped, interacting Bose–Einstein condensates which did not receive much attention until recently is the fact that not only the interaction strength, but also the trap geometry can have a strong influence on the nature of a condensate. While the conceptually simplest condensates are fully condensed, other types of condensates exist. In so called fragmented condensates two or more single-particle quantum states are occupied by a large number of atoms [38, 39]. The concept of fragmentation will be explained in detail in Chap. 2. Fragmented condensates were initially thought to be unphysical [38, 39]. The contrary is the case, especially in double-well and multi-well traps. As it turns out, already the ground state of condensates in such traps is fragmented, provided that the separating potential barrier is high enough [40–48]. When long-range interparticle interactions are present also the ground state of single-well traps can be fragmented [49]. While fully condensed systems behave like classical fields such as coherent light pulses in optical fibers or certain kinds of water waves, fragmented condensates have no simple classical equivalent and cannot be described by Gross–Pitaevskii theory.

Apart from Gross–Pitaevskii theory another very popular approximation for the theoretical description of trapped ultracold bosons is the Bose–Hubbard model [50–53]. We will derive the Bose–Hubbard model from the Schrödinger equation in Chap. 4. Unlike Gross–Pitaevskii theory, the Bose–Hubbard model is capable of describing fragmented condensates. However, it makes explicit use of the trap potential which, strictly speaking, is assumed to be a lattice of potential wells. It is therefore not as generally applicable as Gross–Pitaevskii theory. Within the framework of the Bose–Hubbard model bosons are allowed to move through the lattice by hopping from one lattice site to a neighboring one. By construction it is a spatially discrete lattice model and thereby very different from Gross–Pitaevskii theory.

Both of these models, Gross–Pitaevskii theory and the Bose–Hubbard model are often used to explain experiments. In fact, the literature on Bose–Einstein condensates relies heavily, almost exclusively on these two approximations. However, very little is known about the true physics which is governed by the many-body Schrödinger equation. Very recently, a new, strictly variational many-body method has been developed for bosons [54–58]. The method is known as the Multiconfigurational time-dependent Hartree for bosons, or short MCTDHB. This method allowed for the first time the solution of the time-dependent many-boson Schrödinger equation for large numbers of particles. Thus, it became possible to obtain *exact* results on the many-body dynamics of Bose–Einstein condensates in arbitrary trap geometries from first principles. The thereby obtained results reveal without exception that even for the simplest cases and on short time scales the true physics of interacting many-boson systems is far richer than what can be anticipated based on Gross–Pitaevskii theory or the Bose–Hubbard model, see Refs. [48, 54, 55, 59–67]. Some of these results where obtained in the framework of this thesis. More information about the MCTDHB method can be found at http://MCTDHB.uni-hd.de. An Open-Source implementation of the method can be downloaded from http://OpenMCTDHB.uni-hd.de.

1.3 Overview of this Thesis

This thesis is organized as follows. In Chap. 2 we review the most important concepts of the theory of ultracold bosons. We begin with the many-body Hamiltonian, its different representations and show how the Schrödinger equation can be obtained from a variational principle. The representation of a many-body wave function in a finite basis set and its implications are discussed. We then introduce reduced density matrices, summarize their properties and discuss their relation to observables At the end of the chapter we review the criterion for Bose–Einstein condensation, define fragmented condensates rigorously in terms of reduced density matrices and classify regimes of interacting bosons.

In Chap. 3 we review general methods for Bose–Einstein condensates and give a self-contained derivation of the Multiconfigurational time-dependent Hartree for bosons method. We also show that the celebrated Gross–Pitaevskii equation is contained in this method as a special case.

Chapter 4 discusses the most commonly used lattice models for the quantum dynamics of ultracold bosons as far as they are relevant to this work. In particular these are the two-mode Gross–Pitaevskii model and the Bose–Hubbard model for two sites. A short overview of related topics, such as Wannier functions, self-trapping and commonly employed validity criteria for the Bose–Hubbard model is given.

In Chap. 5 we turn to a specific physical example of a trapped Bose–Einstein condensate and focus on the ground state of one thousand bosons in a one-dimensional double-well potential at different barrier heights. We solve the time-independent many-body Schrödinger equation and monitor how the ground state becomes more and more fragmented with increasing barrier height. This transition manifests itself in the correlation functions and the coherence of the condensate. In the limits of a low and a very high barrier we show that a mean-field description is applicable, while the state is a true many-body state in between those limits. The strongest correlations are shown to occur there.

Chapter 6 is devoted to the dynamics of bosonic Josephson junctions, obtained by solving the time-dependent many-body Schrödinger equation numerically exactly for up to one hundred particles. The exact results are compared to those of the most popular theories of the field, Gross–Pitaevskii theory and of the Bose–Hubbard model. Both of these theories are shown to deviate from the exact results after short times and at weak interaction strengths. Self-trapping is shown to be largely suppressed by the many-body dynamics at weak interactions. For stronger interactions we find self-trapping for some time on the full many-body level and a completely novel equilibration dynamics which is accompanied by a quick loss of coherence. The resulting many-body state is found to be highly correlated and many-fold fragmented.

Chapter 7 extends the study of bosonic Josephson junctions and investigates the dynamics for attractive and repulsive interactions of equal magnitude, studied here by comparing Bose–Hubbard and numerically exact results of the many-body Schrödinger equation. It is shown that a symmetry of the Bose–Hubbard model dictates an equivalence between the evolution in time for attractive and repulsive interactions. The many-body Schrödinger equation does not possess this symmetry and consequently the dynamics of the attractive and repulsive junctions are different.

In Chap. 8 we present a new concept to optimize many of the lattice models of condensed matter physics. In particular we allow the Wannier functions of a lattice model to depend on time, and show how equations of motion for such generalized lattice models can be derived from the variational principle. The concept is general and can be applied to any lattice model that employs Wannier functions, e.g. the Bose–Hubbard model, the Fermi Hubbard model and any extensions or combinations of these. Since all parameters in the model are determined by the variational principle, the so obtained models are optimal. As an example we illustrate this concept for the Bose–Hubbard model. In a quantum quench scenario the results of the conventional Bose–Hubbard model and of the optimal Bose–Hubbard model are compared to numerically exact results of the many-body Schrödinger equation. These exact results are obtained by using MCTDHB and consitute the first numerically exact results on quantum quenches. Thereby it is shown that lattice models can be optimized at little extra computational cost.

References

1. S.N. Bose, Z. Phys. **26**, 178 (1924)
2. A. Einstein, Sitzungsberichte der Preussischen Akademie der Wissenschaften. Physikalisch-mathematische Klasse **22**, 261 (1924)
3. A. Einstein, Sitzungsberichte der Preussischen Akademie der Wissenschaften. Physikalisch-mathematische Klasse **1**, 3 (1925)
4. F. London, Nature **141**, 643 (1938)
5. F. London, Nat. Phys. Rev. **54**, 947 (1938)
6. O. Penrose, L. Onsager, Phys. Rev. **104**, 576 (1956)
7. T.R. Sosnick, W.M. Snow, P.E. Sokol, Phys. Rev. B **41**, 11185 (1990)
8. P.E. Sokol, in *Bose–Einstein Condensation*, ed. by A. Griffin, D.W. Snoke, S. Stringari (Cambridge University Press, Cambridge, 1996)
9. D.M. Ceperly, Rev. Mod. Phys. **67**, 279 (1995)
10. A. Griffin, D.W. Snoke, S. Stringari (eds.), *Bose–Einstein Condensation* (Cambridge University Press, Cambridge, 1996)
11. S. Chu, Rev. Mod. Phys. **70**, 685 (1998)
12. C. Cohen-Tannoudji, Rev. Mod. Phys. **70**, 707 (1998)
13. W.D. Phillips, Rev. Mod. Phys. **70**, 721 (1998)
14. C. Monroe, W. Swann, H. Robinson, C. Wieman, Phys. Rev. Lett. **65**, 1571 (1990)
15. C. Monroe, E. Cornell, C. E. Wieman, in *Proceedings of the International School of Physics "Enrico Fermi", Course CVXIII, Laser Manipulation of Atoms and Ions* (North Holland, Amsterdam, 1992), p. 361
16. M.H. Anderson, J.R. Ensher, M.R. Matthews, C.E. Wieman, E.A. Cornell, Science **269**, 198 (1995)
17. C.C. Bradley, C.A. Sackett, J.J. Tollett, R.G. Hulet, Phys. Rev. Lett. **75**, 1687 (1995)
18. K.B. Davis, M.-O. Mewes, M.R. Andrews, van N.J. Druten, D.S. Durfee, D.M. Kurn, W. Ketterle, Phys. Rev. Lett. **75**, 3969 (1995)
19. *Advanced information on the Nobel Prize in Physics 2001*, The Royal Swedish Academy of Sciences (2001), http://www.nobelprize.org/nobel_prizes/physics/laureates/2001/adv.html
20. C.C. Bradley, C.A. Sackett, R.G. Hulet, Phys. Rev. Lett. **78**, 985 (1997)
21. *Information for the Public, The Nobel Prize in Physics 2001*, The Royal Swedish Academy of Sciences (2001), http://www.nobelprize.org/nobel_prizes/physics/laureates/2001/public.html
22. W. Ketterle, Rev. Mod. Phys. **74**, 1131 (2002)
23. P. Courteille, R. Freeland, D. Heinzen, van F. Abeelen, B. Verhaar, Phys. Rev. Lett. **81**, 69 (1998)

24. S. Inouye, M. Andrews, J. Stenger, H.J. Miesner, S. Stamper-Kurn, W. Ketterle, Nature **392**, 151 (1998)
25. F. Dalfovo, S. Giorgini, L.P. Pitaevskii, S. Stringari, Rev. Mod. Phys. **71**, 463 (1999)
26. A. Leggett, Rev. Mod. Phys. **73**, 307 (2001)
27. O. Morsch, M. Oberthaler, Rev. Mod. Phys. **78**, 179 (2006)
28. I. Bloch, J. Dalibard, W. Zwerger, Rev. Mod. Phys. **80**, 885 (2008)
29. S. Giorgini, L.P. Pitaevskii, S. Stringari, Rev. Mod. Phys. **80**, 1215 (2008)
30. A.L. Fetter, Rev. Mod. Phys. **81**, 647 (2009)
31. F. Schwabl, *Statistical Mechanics*, 2nd edn. (Springer, Berlin, 2006)
32. C.J. Pethick, H. Smith, *Bose–Einstein Condensation in Dilute Gases*, 2nd edn. (Cambridge University Press, Cambridge, 2008)
33. L.V. Hau, B.D. Busch, C. Liu, Z. Dutton, M.M. Burns, J.A. Golovchenko, Phys. Rev. A **58**, R54 (1998)
34. F. Schwabl, *Advanced Quantum Mechanics*, 4th edn. (Springer, Berlin, 2008)
35. E.P. Gross, Nuovo Cimento **20**, 454 (1961)
36. L.P. Pitaevskii, Zh. Eksp. Teor. Fiz. **40**, 646 (1961)
37. L.P. Pitaevskii, Sov. Phys. JETP **13**, 451 (1961)
38. P. Nozières, D. Saint James, J. Phys. (Paris) **43**, 1133 (1982)
39. P. Nozières, in *Bose–Einstein Condensation*, ed. by A. Griffin, D.W. Snoke, S. Stringari (Cambridge University Press, Cambridge, 1996)
40. R.W. Spekkens, J.E. Sipe, Phys. Rev. A **59**, 3868 (1999)
41. L.S. Cederbaum, A.I. Streltsov, Phys. Lett. A **318**, 564 (2003)
42. A.I. Streltsov, L.S. Cederbaum, N. Moiseyev, Phys. Rev. A **70**, 053607 (2004)
43. O.E. Alon, L.S. Cederbaum, Phys. Rev. Lett. **95**, 140402 (2005)
44. O.E. Alon, A.I. Streltsov, L.S. Cederbaum, Phys. Rev. Lett. **95**, 030405 (2005)
45. O.E. Alon, A.I. Streltsov, L.S. Cederbaum, Phys. Lett. A **347**, 88 (2005)
46. A.I. Streltsov, O.E. Alon, L.S. Cederbaum, Phys. Rev. A **73**, 063626 (2006)
47. E.J. Mueller, T.-L. Ho, M. Ueda, G. Baym, Phys. Rev. A **74**, 033612 (2006)
48. K. Sakmann, A.I. Streltsov, O.E. Alon, L.S. Cederbaum, Phys. Rev. A **78**, 023615 (2008)
49. P. Bader, U.R. Fischer, Phys. Rev. Lett. **103**, 060402 (2009)
50. M.P.A. Fisher, P.B. Weichman, G. Grinstein, D.S. Fisher, Phys. Rev. B **40**, 546 (1989)
51. G.J. Milburn, J. Corney, E.M. Wright, D.F. Walls, Phys. Rev. A **55**, 4318 (1997)
52. D. Jaksch, C. Bruder, J.I. Cirac, C.W. Gardiner, P. Zoller, Phys. Rev. Lett. **81**, 3108 (1998)
53. D. Jaksch, P. Zoller, Ann. Phys. **315**, 52 (2005)
54. A.I. Streltsov, O.E. Alon, L.S. Cederbaum, Phys. Rev. Lett. **99**, 030402 (2007)
55. O.E. Alon, A.I. Streltsov, L.S. Cederbaum, Phys. Rev. A **77**, 033613 (2008)
56. A.I. Streltsov, O.E. Alon, L.S. Cederbaum, Phys. Rev. A **81**, 022124 (2010)
57. A.I. Streltsov, K. Sakmann, A.U.J. Lode, O.E. Alon, L.S. Cederbaum, The MCTDHB Package, Version 2.1, Heidelberg, (2011), http://MCTDHB.uni-hd.de
58. K. Sakmann, A.U.J. Lode, A.I. Streltsov, O.E. Alon, L.S. Cederbaum, OpenMCTDHB v2.1 (2011), http://OpenMCTDHB.uni-hd.de
59. A.I. Streltsov, O.E. Alon, L.S. Cederbaum, Phys. Rev. Lett. **100**, 130401 (2008)
60. O.E. Alon, A.I. Streltsov, L.S. Cederbaum, Phys. Lett. A **373**, 301 (2009)
61. J. Grond, J. Schmiedmayer, U. Hohenester, Phys. Rev. A **79**, 021603(R) (2009)
62. A.I. Streltsov, O.E. Alon, L.S. Cederbaum, Phys. Rev. A **80**, 043616 (2009)
63. K. Sakmann, A.I. Streltsov, O.E. Alon, L.S. Cederbaum, Phys. Rev. Lett. **103**, 220601 (2009)
64. J. Grond, von G. Winckel, J. Schmiedmayer, U. Hohenester, Phys. Rev. A **80**, 053625 (2009)
65. A.I. Streltsov, K. Sakmann, O.E. Alon, L.S. Cederbaum, Phys. Rev. A **83**, 043604 (2011)
66. K. Sakmann, A.I. Streltsov, O.E. Alon, L.S. Cederbaum, Phys. Rev A **82**, 013620 (2010)
67. J. Grond, J. Schmiedmayer, U. Hohenester, New J. Phys. **12**, 065036 (2010)

Chapter 2
General Theory

In this chapter we review the most important concepts of the theory of ultracold bosons. We begin with the many-body Hamiltonian, its different representations and show how the Schrödinger equation can be obtained from to the time-dependent variational principle. The representation of a many-body wave function in a finite basis set and its implications are discussed. We then introduce reduced density matrices, summarize their properties and discuss their relation to observables. At the end of the Chapter we review the criteria for Bose–Einstein condensation and classify regimes of interacting bosons. We assume that the reader is familiar with the formalism of second quantization, as treated, e.g. in the first three chapters of Ref. [1].

2.1 The Field Operator

All theories and models used in this work can be cast into the same framework if the (spinless) bosonic Schrödinger picture field operator $\hat{\mathbf{\Psi}}(\mathbf{r})$ is taken as a starting point. Here, we work in D-dimensions. In one-dimension (1D) we set $\mathbf{r}=x$. $\hat{\mathbf{\Psi}}(\mathbf{r})$ satisfies the usual bosonic commutation relations

$$\left[\hat{\mathbf{\Psi}}(\mathbf{r}), \hat{\mathbf{\Psi}}^\dagger(\mathbf{r}')\right] = \delta(\mathbf{r}-\mathbf{r}'), \quad \left[\hat{\mathbf{\Psi}}(\mathbf{r}), \hat{\mathbf{\Psi}}(\mathbf{r}')\right] = 0. \tag{2.1}$$

It is convenient to expand $\hat{\mathbf{\Psi}}(\mathbf{r})$ in a *complete* set of orthonormal single-particle functions $\{\phi_k\} = \{\phi_1, \phi_2, \phi_3, \ldots\}$, which we will call *orbitals* for brevity. The orbitals are usually taken as a set of previously known, time-independent functions, e.g. plane waves or harmonic oscillator eigenfunctions. However, it is important to realize that in the most general case neither does the set $\{\phi_k\}$ have to be known, nor is it necessary that the orbitals be time-independent. In the following we will assume that the orbitals $\{\phi_k(\mathbf{r}, t)\}$ are *time-dependent* and form a complete orthonormal set at all times

$$\langle\phi_k|\phi_j\rangle = \delta_{kj}. \tag{2.2}$$

K. Sakmann, *Many-Body Schrödinger Dynamics of Bose–Einstein Condensates*, Springer Theses, DOI: 10.1007/978-3-642-22866-7_2,

Of course this includes also the special case where the orbitals are time-independent: $\phi_k(\mathbf{r}, t) = \phi_k(\mathbf{r})$. The expansion of the time-independent field operator $\hat{\Psi}(\mathbf{r})$ in the time-dependent basis set $\{\phi_k(\mathbf{r}, t)\}$ then reads

$$\hat{\Psi}(\mathbf{r}) = \sum_{k=1}^{\infty} \hat{b}_k(t)\phi_k(\mathbf{r}, t). \tag{2.3}$$

The time-dependent annihilation and creation operators $b_k(t)$ and $b_j^{\dagger}(t)$ obey the commutation relations

$$\left[\hat{b}_k(t), \hat{b}_j^{\dagger}(t)\right] = \delta_{kj} \tag{2.4}$$

for bosons at any time. Substituting Eq. 2.3 into the first of Eq. 2.1 results in the closure relation

$$\sum_{k=1}^{\infty} \phi_k(\mathbf{r}, t)\phi_k^*(\mathbf{r}', t) = \delta(\mathbf{r} - \mathbf{r}'), \tag{2.5}$$

which expresses the fact that the time-dependent set of orbitals $\{\phi_k(\mathbf{r}, t)\}$ is complete.

2.2 The Many-Body Hamiltonian

The Hamiltonian of a system of N spinless particles interacting via pairwise interactions is given by

$$H = \sum_{i}^{N} h(\mathbf{r}_i) + \sum_{i<j}^{N} W(\mathbf{r}_i - \mathbf{r}_j). \tag{2.6}$$

Here

$$h(\mathbf{r}) = -\frac{1}{2}\frac{\partial^2}{\partial \mathbf{r}^2} + V(\mathbf{r}) \tag{2.7}$$

is a one-body Hamiltonian consisting of a kinetic part $T(\mathbf{r})$ and an external potential $V(\mathbf{r})$. We work in dimensionless units where $\hbar = m = 1$. The connection to a real physical Hamiltonian will be made whenever appropriate. $W(\mathbf{r} - \mathbf{r}')$ is the two-body interaction potential. The Hamiltonian H can also be explicitly time-dependent, e.g. because of a time-dependent one-particle potential $V(\mathbf{r}, t)$, but here we restrict the discussion to the case where H is time-independent. In second quantized form the many-body Hamiltonian (2.6) then reads

$$\hat{\mathcal{H}} = \int d\mathbf{r} \hat{\Psi}^{\dagger}(\mathbf{r}) h(\mathbf{r}) \hat{\Psi}(\mathbf{r}) + \frac{1}{2}\int d\mathbf{r} \int d\mathbf{r}' \hat{\Psi}^{\dagger}(\mathbf{r}) \hat{\Psi}^{\dagger}(\mathbf{r}') W(\mathbf{r} - \mathbf{r}') \hat{\Psi}(\mathbf{r}) \hat{\Psi}(\mathbf{r}') \tag{2.8}$$

or equivalently after substituting the representation for the field operator given in Eq. 2.3

$$\hat{H} = \sum_{k,q} \hat{b}_k^\dagger(t)\hat{b}_q(t)h_{kq}(t) + \frac{1}{2}\sum_{k,s,l,q} \hat{b}_k^\dagger(t)\hat{b}_s^\dagger(t)\hat{b}_l(t)\hat{b}_q(t)W_{ksql}(t). \tag{2.9}$$

Here, the matrix elements of $h(\mathbf{r})$ and $W(\mathbf{r}-\mathbf{r}')$ are given by

$$\begin{aligned} h_{kq}(t) &= \int d\mathbf{r}\phi_k^*(\mathbf{r},t)h(s\mathbf{r})\phi_q(\mathbf{r},t), \\ W_{ksql}(t) &= \int d\mathbf{r}\int d\mathbf{r}'\phi_k^*(\mathbf{r},t)\phi_s^*(\mathbf{r}',t)W(\mathbf{r}-\mathbf{r}')\phi_q(\mathbf{r},t)\phi_l(\mathbf{r}',t). \end{aligned} \tag{2.10}$$

If the set of orbitals $\{\phi_k\}$ is time-independent the matrix elements (2.10) are also time-independent and only have to be evaluated once in a computation. However, for time-dependent orbitals the matrix elements (2.10) have to be evaluated at every time step of a computation. In fact, the evaluation of the matrix elements (2.10) can become the performance limiting factor then, especially if computations are done in two and three dimensions. In Appendix A we present an algorithm for the efficient evaluation of the interaction matrix elements $W_{ksql}(t)$ and the one-body matrix elements $h_{kq}(t)$. In both cases the advantages of the fast Fourier transform can be exploited.

2.3 The Time-Dependent Variational Principle

The physical laws of optics, classical mechanics, quantum mechanics, electrodynamics, general relativity and elementary particle physics can be derived from variational principles. While the time-independent variational principle in quantum mechanics can be found in virtually any introductory textbook, the time-dependent version of it is less known. In this section we review the time-dependent variational principle for quantum mechanics, on which all of the numerical methods used in this work can be based. We follow the exposition of Ref. [2] and start from the Lagrangian

$$\mathcal{L}[\Psi(t), \Psi^*(t)] = \left\langle \Psi(t) \left| \hat{H} - i\frac{\partial}{\partial t} \right| \Psi(t) \right\rangle, \tag{2.11}$$

where the wave function $\Psi(t)$ is required to be normalized at all times $\langle\Psi(t)|\Psi(t)\rangle = 1$. We write $\dot{\Psi} \equiv \frac{\partial}{\partial t}\Psi(t)$ for the time derivative. The equations of motion are then determined by the principle of least action $\delta S = 0$. Using $\langle\Psi|\dot{\Psi}\rangle = -\langle\dot{\Psi}|\Psi\rangle$ it can be verified that arbitrary variations of the actionfunctional

$$S[\Psi, \Psi^*] = \int_{t_0}^{t_1} dt' \mathcal{L}[\Psi(t'), \Psi^*(t')] \tag{2.12}$$

with respect to $\langle\Psi|$ and $|\Psi\rangle$ yield the Schrödinger equation and its hermitian conjugate

$$\begin{aligned} \hat{H}|\Psi\rangle &= i|\dot{\Psi}\rangle \\ \langle\Psi|\hat{H} &= -i\langle\dot{\Psi}|, \end{aligned} \tag{2.13}$$

where we have assumed a hermitian Hamiltonian $\hat{H}^\dagger = \hat{H}$. In all practical computations the variations will not be taken with respect to the wave function itself, but rather an *ansatz* for the wave function must be made, containing parameters. The variation is then taken with respect to these parameters. By including more and more parameters into the *ansatz* wave function, the accuracy can improved successively until convergence is reached.

2.4 The Many-Boson Wave Function

It is a well known fact that any wave function of identical fermions can be expanded in a complete set of Slater determinants. The bosonic equivalent of a Slater determinant is called a *permanent*. In this work we are dealing with bosons only, and we will also use the term *configuration* interchangeably with the term permanent. For a given set of M time-dependent orbitals $\{\phi_1, \ldots, \phi_M\}$ and N bosons a total of $\binom{N+M-1}{N}$ permanents

$$|n_1, n_2, \ldots, n_M; t\rangle = \frac{1}{\sqrt{n_1! n_2! \cdots n_M!}} \left(\hat{b}_1^\dagger(t)\right)^{n_1} \left(\hat{b}_2^\dagger(t)\right)^{n_2} \cdots \left(\hat{b}_M^\dagger(t)\right)^{n_M} |vac\rangle \tag{2.14}$$

can be constructed by distributing the N bosons over the M orbitals. We collect the occupations in the vector $\vec{n} = (n_1, n_2, \ldots, n_M)$, where $n_1 + n_2 + \cdots + n_M = N$. The most general *ansatz* for the many-body wave function $|\Psi(t)\rangle$ of N identical bosons is a linear combination of such time-dependent permanents

$$|\Psi(t)\rangle = \sum_{\vec{n}} C_{\vec{n}}(t) \, |n_1, n_2, \ldots, n_M; t\rangle, \tag{2.15}$$

where the summation in (2.15) runs over all $\binom{N+M-1}{N}$ permanents. Of course, if M goes to infinity then the *ansatz* (2.15) for the wave function becomes exact since the set of permanents $|n_1, n_2, \ldots, n_M; t\rangle$ spans the complete N-particle Hilbert space.

In all practical computations M is finite and the set of orbitals $\{\phi_1, \ldots, \phi_M\}$ only approximates a complete set. Using the orbitals $\{\phi_1, \ldots, \phi_M\}$ as an approximation to a complete set, the finite size representations of the field operator $\hat{\boldsymbol{\Psi}}_M$ and the closure relation read

$$\hat{\boldsymbol{\Psi}}_M(\mathbf{r};t) = \sum_{k=1}^{M} \hat{b}_k(t)\phi_k(\mathbf{r},t). \tag{2.16}$$

and

$$\sum_{k=1}^{M} \phi_k(\mathbf{r},t)\phi_k^*(\mathbf{r}',t) = \delta_M(\mathbf{r}-\mathbf{r}';t), \tag{2.17}$$

both of which are generally time-dependent. Here $\delta_M(\mathbf{r}-\mathbf{r}';t)$ is the finite size approximation of a true delta function. For any function that lies within the Hilbert space spanned by the orbitals $\{\phi_1,\ldots,\phi_M\}$, $\hat{\boldsymbol{\Psi}}_M$ and δ_M act like their exact equivalents, Eqs. 2.3 and 2.5. For any finite M *ansatz* for the wave function the finite M representation of the many-body Hamiltonian can be obtained by substituting the respective finite M expansion for the field operator, Eq. 2.16 into Eq. 2.8. This allows for a comparison of all methods discussed in this work on the same footing.

We note that allowing the orbitals $\{\phi_1,\ldots,\phi_M\}$ to depend on time is a generalization of the conventional expansion in a time-independent basis set that does not lead to any additional complications as far as quantities at a single time t are concerned. In principle any time-dependent set of orthonormal orbitals $\{\phi_1,\ldots,\phi_M\}$ could be chosen. However, this additional freedom is most effectively used, if the orbitals are determined by the time-dependent variational principle [2]. In Sect. 3.1 we discuss the Multiconfigurational time-dependent Hartree for bosons method (MCTDHB), which does exactly that. If the orbitals $\{\phi_1,\ldots,\phi_M\}$ are not allowed to depend on time, only the coefficients $C_{\vec{n}}(t)$ in the ansatz wave function (2.15) remain time-dependent and Eqs. 2.16 and 2.17 constitute time-independent approximations of the exact field operator, Eq. 2.3, and the exact closure relation, Eq. 2.5.

The *ansatz* (2.15) is the most general M-orbital many-boson *ansatz* possible. Less general *ansatz* wave functions are common in the literature. In the theory of many-particle systems the following distinction is made based on the form of the *ansatz* wave function. An *ansatz* for a many-boson wave function is a

- mean-field *ansatz*, if it consists of a single permanent, or a
- many-body *ansatz* otherwise.

The MCTDHB method mentioned above uses the most general M-orbital ansatz (2.15) and is therefore a many-body method. If only $M=1$ orbital is used in Eq. 2.15 the *ansatz* wave function becomes a single permanent, i.e. a mean-field *ansatz*, with all N particles in one orbital. The celebrated Gross-Pitaevskii theory discussed in Sect. 3.2 is of this type. In the two-mode Gross-Pitaevskii model, discussed in Sect. 4.2, a restricted *ansatz* is made for the orbital of Gross-Pitaevskii theory. The model is therefore a mean-field model that approximates Gross-Pitaevskii theory. Lastly, the Bose-Hubbard model, discussed in Sect. 4.3, uses a many-body ansatz for the wave function, but unlike in MCTDHB the orbitals are not allowed to depend on time.

2.5 Reduced Density Matrices and their Eigenfunctions

We consider a given wave function $\Psi(\mathbf{r}_1,\ldots,\mathbf{r}_N;t)$ of N identical, spinless bosons with spatial coordinates $\mathbf{r}_i$. The pth order reduced density matrix (RDM), is defined by

$$\rho^{(p)}(\mathbf{r}_1,\ldots,\mathbf{r}_p|\mathbf{r}'_1,\ldots,\mathbf{r}'_p;t)=\frac{N!}{(N-p)!}\int\Psi(\mathbf{r}_1,\ldots,\mathbf{r}_p,\mathbf{r}_{p+1},\ldots,\mathbf{r}_N;t)\\ \times\Psi^*(\mathbf{r}'_1,\ldots,\mathbf{r}'_p,\mathbf{r}_{p+1},\ldots,\mathbf{r}_N;t)d\mathbf{r}_{p+1}\ldots d\mathbf{r}_N, \tag{2.18}$$

where the wave function is assumed to be normalized $\langle\Psi(t)|\Psi(t)\rangle=1$. The pth order RDM, Eq. 2.18, can be regarded as the kernel of the operator

$$\hat{\rho}^{(p)}=\frac{N!}{(N-p)!}\mathrm{Tr}_{N-p}\Big[|\Psi(t)\rangle\langle\Psi(t)|\Big] \tag{2.19}$$

in Fock space, where $\mathrm{Tr}_{N-p}[\cdot]$ specifies taking the partial trace over $N-p$ particles. Since the wave function is symmetric in its coordinates, it does not matter over which particles the trace is taken. In what follows, we add $|\Psi\rangle$ as an additional subscript if a result is only valid for states $|\Psi\rangle$ of a particular form. The diagonal $\rho^{(p)}(\mathbf{r}_1,\ldots,\mathbf{r}_p|\mathbf{r}_1,\ldots,\mathbf{r}_p;t)$ is the p-particle probability distribution at time t normalized to $N!/(N-p)!$.

In order to discuss observables not only in real space, but also in momentum space, we define the Fourier transform of a function $f(\mathbf{r}_1,\ldots,\mathbf{r}_p)$ of p D-dimensional coordinates $\mathbf{r}_i$ by

$$f(\mathbf{k}_1,\ldots,\mathbf{k}_p)=\frac{1}{(2\pi)^{pD/2}}\int d^p\mathbf{r}e^{-i\sum_{l=1}^{p}k_l r_l}f(\mathbf{r}_1,\ldots,\mathbf{r}_p). \tag{2.20}$$

By applying the Fourier transform, Eq. 2.20, to the $2p$ coordinates $\mathbf{r}_1,\ldots,\mathbf{r}_p$ and $\mathbf{r}'_1,\ldots,\mathbf{r}'_p$ in Eq. 2.18, the momentum space representation of $\hat{\rho}^{(p)}$:

$$\rho^{(p)}(\mathbf{k}_1,\ldots,\mathbf{k}_p|\mathbf{k}'_1,\ldots,\mathbf{k}'_p;t)=\frac{N!}{(N-p)!}\int\Psi(\mathbf{k}_1,\ldots,\mathbf{k}_p,\mathbf{k}_{p+1},\ldots,\mathbf{k}_N;t)\\ \times\Psi^*(\mathbf{k}'_1,\ldots,\mathbf{k}'_p,\mathbf{k}_{p+1},\ldots,\mathbf{k}_N;t)d\mathbf{k}_{p+1}\ldots d\mathbf{k}_N \tag{2.21}$$

is obtained. The diagonal $\rho^{(p)}(\mathbf{k}_1,\ldots,\mathbf{k}_p|\mathbf{k}_1,\ldots,\mathbf{k}_p;t)$ in momentum space is the p-particle momentum distribution at time t, normalized to $N!/(N-p)!$.

The pth order RDM $\rho^{(p)}$ can be expanded in its eigenfunctions, leading to the representations

$$\rho^{(p)}(\mathbf{r}_1,\ldots,\mathbf{r}_p|\mathbf{r}'_1,\ldots,\mathbf{r}'_p;t)=\sum_i n_i^{(p)}(t)\alpha_i^{(p)}(\mathbf{r}_1,\ldots,\mathbf{r}_p,t)\alpha_i^{(p)*}(\mathbf{r}'_1,\ldots,\mathbf{r}'_p,t) \tag{2.22}$$

and

$$\rho^{(p)}(\mathbf{k}_1,\ldots,\mathbf{k}_p|\mathbf{k}'_1,\ldots,\mathbf{k}'_p;t)=\sum_i n_i^{(p)}(t)\alpha_i^{(p)}(\mathbf{k}_1,\ldots,\mathbf{k}_p,t)\alpha_i^{(p)*}(\mathbf{k}'_1,\ldots,\mathbf{k}'_p,t). \tag{2.23}$$

Here, $n_i^{(p)}(t)$ denotes the ith eigenvalue of the pth order RDM and $\alpha_i^{(p)}$ the corresponding (symmetric) eigenfunction in real space. The momentum space representation can be obtained by applying the Fourier transform to the real space eigenfunctions. All eigenfunctions $\alpha_i^{(p)}$ are normalized to one. The eigenfunctions are known as *natural p-functions* and the eigenvalues as *natural occupations*. For $p=1$ and $p=2$ the eigenfunctions are known as natural orbitals and natural geminals, respectively. We order the eigenvalues $n_i^{(p)}(t)$ for every p non-increasingly, such that $n_1^{(p)}(t)$ denotes the largest eigenvalue of the pth order RDM. The normalization of the many-body wave function and Eqs. 2.18, 2.22 and 2.23 put the restriction

$$\sum_i n_i^{(p)}(t)=\frac{N!}{(N-p)!} \tag{2.24}$$

on the eigenvalues of the pth order RDM. Thus the largest eigenvalue $n_1^{(p)}(t)$ is bounded from above by [3, 4]

$$n_1^{(p)}(t)\leq\frac{N!}{(N-p)!}. \tag{2.25}$$

Lower bounds on $n_1^{(p)}(t)$ can be derived, relating RDMs of different order [5, 6].

2.6 Two-Body Expectation Values and Geminals

For a system of identical particles interacting via two-body interactions only it is possible to express the expectation value of the many-body Hamiltonian (or any other two-body operator) by an expression involving only the second order RDM of the system. Consider a general many-body Hamiltonian of the type given in Eq. 2.6, i.e. consisting of one-body operators $h(\mathbf{r}_i)$ and two-body operators $W(\mathbf{r}_i-\mathbf{r}_j)$. It is then useful to define the *reduced Hamiltonian operator*

$$K(\mathbf{r}_1,\mathbf{r}_2)=h(\mathbf{r}_1)+h(\mathbf{r}_2)+(N-1)W(\mathbf{r}_1-\mathbf{r}_2). \tag{2.26}$$

Following Refs. [3, 4] and making use of the time-dependent natural geminals $\alpha_i^{(2)}(\mathbf{r}_1,\mathbf{r}_2,t)$ the expectation value of the energy E can then be expressed at any time through the equation [7]:

$$E = \frac{1}{2} N \sum_i n_i^{(2)}(t) \int d\mathbf{r}_1 d\mathbf{r}_2 \alpha_i^{(2)^*}(\mathbf{r}_1, \mathbf{r}_2, t) K(\mathbf{r}_1, \mathbf{r}_2) \alpha_i^{(2)}(\mathbf{r}_1, \mathbf{r}_2, t). \quad (2.27)$$

The expression (2.27) is usually written for time-independent states, but it holds equally well in the dynamical case [7]. Note that the many-body wave function does not appear explicitly in Eq. 2.27. In order to determine the energy spectrum of a system of identical particles it is therefore tempting to devise a variational procedure that minimizes Eq. 2.27 by variations over the Hilbert space of two-particle functions. However, the requirement of a totally symmetric (or antisymmetric) many-body wave function puts an additional constraint on the space of allowed two-particle functions which is not easy to satisfy. This is known as the *N*-representability problem [3, 4]. The determination of tight bounds on eigenvalues of RDMs is therefore an active field of research. We will not go any further into the details of these approaches to many-body physics and refer the reader to the literature [3, 4].

2.7 Correlation Functions and RDMs

Equivalent to Eq. 2.18, the pth order RDM can be defined using field operators as

$$\rho^{(p)}(\mathbf{r}_1, \ldots, \mathbf{r}_p | \mathbf{r}'_1, \ldots, \mathbf{r}'_p; t) = \langle \Psi(t) | \hat{\boldsymbol{\Psi}}^\dagger(\mathbf{r}'_1) \ldots \hat{\boldsymbol{\Psi}}^\dagger(\mathbf{r}'_p) \hat{\boldsymbol{\Psi}}(\mathbf{r}_p) \ldots \hat{\boldsymbol{\Psi}}(\mathbf{r}_1) | \Psi(t) \rangle \quad (2.28)$$

where the Schrödinger field operators satisfy the usual bosonic commutation relations Eq. 2.1. The representation given in Eq. 2.28 shows that the pth order RDM is identical to the pth order *correlation function* at equal times [7–9]. Here we only consider spatial correlations, i.e. correlations at equal times. For correlations at different times there is no equivalence like Eq. 2.28 between correlation functions and RDMs.

2.8 Definition and Signatures of *n*th Order Coherence

Apart from the p-particle distributions themselves, either in real space or in momentum space, it is also of great interest to compare the p-particle probabilities to their respective one-particle probabilities. Thereby, it becomes possible to identify effects that are due to the quantum statistics of the particles. The *normalized pth order correlation function* of an *N*-boson system at time t is defined by [8]

$$g^{(p)}(\mathbf{r}'_1, \ldots, \mathbf{r}'_p, \mathbf{r}_1, \ldots, \mathbf{r}_p; t) = \frac{\rho^{(p)}(\mathbf{r}_1, \ldots, \mathbf{r}_p | \mathbf{r}'_1, \ldots, \mathbf{r}'_p; t)}{\sqrt{\prod_{i=1}^{p} \rho^{(1)}(\mathbf{r}_i | \mathbf{r}_i; t) \rho^{(1)}(\mathbf{r}'_i | \mathbf{r}'_i; t)}} \quad (2.29)$$

and is the key quantity in the definition of spatial coherence.

Full spatial nth order coherence is only obtained if $\rho^{(p)}$ factorizes *for all* $p \leq n$ into a product of *one* complex valued function $\mathcal{E}(\mathbf{r}, t)$ of the form

$$\rho^{(p)}(\mathbf{r}_1,\ldots,\mathbf{r}_p|\mathbf{r}'_1,\ldots,\mathbf{r}'_p;t)=\mathcal{E}^*(\mathbf{r}'_1,t)\cdots\mathcal{E}^*(\mathbf{r}'_p,t)\mathcal{E}(\mathbf{r}_p,t)\cdots\mathcal{E}(\mathbf{r}_1,t). \tag{2.30}$$

In this case one finds $|g^{(p)}|=1$ for all $p\leq n$ by substituting Eq. 2.30 into Eq. 2.29. For $n>1$ Eq. 2.30 cannot be satisfied exactly by any *N*-particle state. However, if all *N* bosons occupy the same orbital a maximally coherent *N*-particle state is obtained and

$$|g^{(p)}(\mathbf{r}'_1,\ldots,\mathbf{r}'_p,\mathbf{r}_1,\ldots,\mathbf{r}_p;t)|=\frac{N!}{(N-p)!N^p} \tag{2.31}$$

for all $p\leq N$ [10]. Independent of the degree of coherence $g^{(p)}=0$ for $p>N$. We note that the absolute value of $g^{(1)}$ is always bounded from above by one, but there is no upper bound on $g^{(p)}$ if $p>1$.

The diagonal $g^{(p)}(\mathbf{r}_1,\ldots,\mathbf{r}_p,\mathbf{r}_1,\ldots,\mathbf{r}_p;t)$ of the normalized *p*th order correlation function in real space is a measure for the degree of *p*th order coherence. The simultaneous detection probabilities at positions $\mathbf{r}_1,\ldots,\mathbf{r}_p$ are correlated (anticorrelated) if the real space diagonal of $g^{(p)}$ is greater (less) than one.

Note that if Eq. 2.30 holds in real space, it also holds in momentum space, as can be seen by Fourier transforming each of the $2p$ variables in Eq. 2.30. It is therefore possible to define the normalized *p*th order correlation function in momentum space by

$$g^{(p)}(\mathbf{k}'_1,\ldots,\mathbf{k}'_p,\mathbf{k}_1,\ldots,\mathbf{k}_p;t)=\frac{\rho^{(p)}(\mathbf{k}_1,\ldots,\mathbf{k}_p|\mathbf{k}'_1,\ldots,\mathbf{k}'_p;t)}{\sqrt{\prod_{i=1}^{p}\rho^{(1)}(\mathbf{k}_i|k_i;t)\rho^{(1)}(\mathbf{k}'_i|\mathbf{k}'_i;t)}}. \tag{2.32}$$

Equivalent to the situation in real space $g^{(p)}(\mathbf{k}_1,\ldots,\mathbf{k}_p,\mathbf{k}_1,\ldots,\mathbf{k}_p;t)$, the diagonal of Eq. 2.27, expresses the tendency of p momenta to be measured simultaneously. The simultaneous detection probabilities of momenta $\mathbf{k}_1,\ldots,\mathbf{k}_p$ are correlated (anticorrelated) if the momentum space diagonal of $g^{(p)}$ is greater (less) than one. We note that the *p*th order momentum distribution, the diagonal of $\rho^{(p)}$ in momentum space, depends on the entire *p*th order RDM $\rho^{(p)}(\mathbf{r}_1,\ldots,\mathbf{r}_p|\mathbf{r}'_1,\ldots,\mathbf{r}'_p;t)$, see Appendix B. Thus, the diagonal of $g^{(p)}$ in momentum space provides information about the coherence of $|\Psi(t)\rangle$ which is not contained in the diagonal of $g^{(p)}$ in real space and vice versa.

2.9 First and Second Order RDMs, Correlations and Coherence

The fact that many-body quantum systems interact generally via two-body interaction potentials makes the RDMs of first and second order particularly important. For example, the expectation value of any two-body operator can be represented by an integral involving the second-order RDM only, see Sect. 2.6. In this section we summarize the most important results about the RDMs of first and second order. For the first order RDM the following set of equations holds:

$$\begin{aligned}\rho^{(1)}(\mathbf{r}_1|\mathbf{r}_1';t) &= N\int d\mathbf{r}_2 d\mathbf{r}_3\cdots d\mathbf{r}_N \Psi(\mathbf{r}_1,\mathbf{r}_2,\ldots,\mathbf{r}_N;t)\Psi^*(\mathbf{r}_1',\mathbf{r}_2,\ldots,\mathbf{r}_N;t)\\ &= \left\langle \Psi(t)\left|\hat{\boldsymbol{\Psi}}^\dagger(\mathbf{r}_1')\hat{\boldsymbol{\Psi}}(r_1)\right|\Psi(t)\right\rangle\\ &= \sum_{k,q}\rho_{kq}(t)\phi_k^*(\mathbf{r}_1',t)\phi_q(\mathbf{r}_1,t),\end{aligned} \tag{2.33}$$

where expressions for the matrix elements of the one-body density matrix $\rho_{kq}(t) = \left\langle\Psi\left|\hat{b}_k^\dagger\hat{b}_q\right|\Psi\right\rangle$ for bosons [11] are listed in Appendix C. Similarly, the reduced two-body density matrix of $\Psi(t)$ is given by

$$\begin{aligned}\rho^{(2)}(\mathbf{r}_1,\mathbf{r}_2|\mathbf{r}_1',\mathbf{r}_2';t) &= N(N-1)\int d\mathbf{r}_3\ldots d\mathbf{r}_N\Psi(\mathbf{r}_1,\mathbf{r}_2,\mathbf{r}_{p+1},\ldots,\mathbf{r}_N;t)\\ &\quad\times\Psi^*(\mathbf{r}_1',\mathbf{r}_2',\mathbf{r}_3,\ldots,\mathbf{r}_N;t)\\ &= \left\langle\Psi(t)\left|\hat{\Psi}^\dagger(\mathbf{r}_1')\hat{\Psi}^\dagger(\mathbf{r}_2')\hat{\Psi}(\mathbf{r}_2)\hat{\Psi}(\mathbf{r}_1)\right|\Psi(t)\right\rangle\\ &= \sum_{k,s,l,q}\rho_{kslq}(t)\phi_k^*(\mathbf{r}_1',t)\phi_s^*(\mathbf{r}_2',t)\phi_l(\mathbf{r}_2,t)\phi_q(\mathbf{r}_1,t),\end{aligned} \tag{2.34}$$

where the matrix elements of the two-body density matrix $\rho_{kslq}(t) = \left\langle\Psi\left|\hat{b}_k^\dagger\hat{b}_s^\dagger\hat{b}_l\hat{b}_q\right|\Psi\right\rangle$ for bosons [11] are collected in Appendix C.

The upper bound on the largest eigenvalue of the pth order RDM, Eq. 2.25 reduces for $p=1$ and $p=2$ to

$$\begin{aligned} n_1^{(1)}(t) &\le N\\ n_1^{(2)}(t) &\le N(N-1)\end{aligned} \tag{2.35}$$

A lower bound on the largest natural geminal occupation number is given by [5]

$$n_1^{(2)}(t) \ge n_1^{(1)}(t)[n_1^{(1)}(t)-1]. \tag{2.36}$$

The inequality

$$N(N-1) \ge n_1^{(2)}(t) \ge n_1^{(1)}(t)[n_1^{(1)}(t)-1]. \tag{2.37}$$

thus bounds the largest natural geminal occupation number from below and from above. The quantities $\rho(\mathbf{r};t) \equiv \rho^{(1)}(\mathbf{r}|\mathbf{r};t)$ and $\rho(\mathbf{k};t) \equiv \rho^{(1)}(\mathbf{k}|\mathbf{k};t)$ are known as the one-particle density and the one-particle momentum distribution, respectively. We remind the reader that these quantities are normalized to N. Similarly, $\rho^{(2)}(\mathbf{r},\mathbf{r}'|\mathbf{r},\mathbf{r}';t)$ and $\rho^{(2)}(\mathbf{k}_1,\mathbf{k}_2|\mathbf{k}_1,\mathbf{k}_2;t)$ are known as the two-particle density and the two-particle momentum distribution, respectively. Both are normalized to $N(N-1)$.

A maximally coherent *N*-boson state is a state in which *all N* particles occupy the same orbital and satisfies

$$\begin{aligned}|g^{(1)}(\mathbf{r}'_1, \mathbf{r}_1; t)| &= |g^{(1)}(\mathbf{k}'_1, \mathbf{k}_1; t)| = 1\\ |g^{(2)}(\mathbf{r}'_1, \mathbf{r}'_2, \mathbf{r}_1, \mathbf{r}_2; t)| &= |g^{(2)}(\mathbf{k}'_1, \mathbf{k}'_2, \mathbf{k}_1, \mathbf{k}_2; t)| = 1 - 1/N,\end{aligned} \tag{2.38}$$

see Eq. 2.31. First order coherence is mathematically defined as the factorization of $g^{(1)}$ into a product of a single function, see Eq. 2.30. The absolute value of $g^{(1)}$ can be determined experimentally by Young double slit experiments under the following assumptions. If I_{max} (I_{min}) is the maximal (minimal) measured intensity on a screen behind a double slit with the slits located at positions $\mathbf{r}'_1$ and $\mathbf{r}_1$, then

$$v = \frac{I_{max} - I_{min}}{I_{max} + I_{min}} \tag{2.39}$$

is known as the *fringe visibility*. If the bosons are noninteracting and if the slits are illuminated with equal intensity then the fringe visibility is given by $v = |g^{(1)}(\mathbf{r}'_1, \mathbf{r}_1; t_0)|$. Here, t_0 is the time of release from the slits. More details on the relation between the fringe visibility and $g^{(1)}$ can be found in textbooks on quantum optics, e.g. Ref. [12]. Apart from photons such double slit experiments have also been done using Bose–Einstein condensates, see e.g. Ref. [13]. However, if interactions during the expansion behind the slit are not negligible, there is no simple relation between the fringe visibility and $g^{(1)}$. In other words, the interaction between the particles can modify the observed interference pattern [14–16].

In order to determine the degree of coherence of a given system, it is necessary to quantify how well Eq. 2.30 is satisfied. However, Eq. 2.30 is a mathematical definition that cannot easily be determined from measurements. However, the quantities $|g^{(1)}(\mathbf{r}'_1, \mathbf{r}_1; t_0)|^2$, $g^{(2)}(\mathbf{r}_1, \mathbf{r}_2, \mathbf{r}_1, \mathbf{r}_2; t)$ and $g^{(2)}(\mathbf{k}_1, \mathbf{k}_2, \mathbf{k}_1, \mathbf{k}_2; t)$ are accessible in experiments. The last two are the diagonals of $g^{(2)}$ in real and in momentum space and can be computed as ratios of expectation values. Deviations from their maximally obtainable values (2.38) are a measure for the coherence of a quantum state. We note that $g^{(1)}(\mathbf{k}'_1, \mathbf{k}_1; t_0)$ does not contain any additional information about the coherence of a state that goes beyond what can be concluded from $g^{(1)}(\mathbf{r}'_1, \mathbf{r}_1; t_0)$ alone. The reverse is true for $g^{(2)}(\mathbf{r}_1, \mathbf{r}_2, \mathbf{r}_1, \mathbf{r}_2; t)$ and $g^{(2)}(\mathbf{k}_1, \mathbf{k}_2, \mathbf{k}_1, \mathbf{k}_2; t)$.

A visualization of the degree of coherence is highly desirable, as it helps to understand the coherence limiting factors in an intuitive manner. In one-dimensional systems $|g^{(1)}(\mathbf{r}'_1, \mathbf{r}_1; t)|^2$ can be represented as a two-dimensional plot. Similarly, the diagonals of $g^{(2)}$ in real and in momentum space can be represented as two-dimensional plots. In Chaps. 5 and 6 we will visualize the degree of first- and second-order coherence of a one-dimensional system.

2.10 Definition of Bose–Einstein Condensation and Fragmentation

The natural orbital occupation numbers $n_i^{(1)}$ serve to define Bose–Einstein condensation. According to Penrose and Onsager [17], a system of identical bosons is said to be *condensed*, if the largest eigenvalue of the first-order RDM is of the order of the number of particles in the system, i.e.

$$n_1^{(1)} = \mathcal{O}(N). \tag{2.40}$$

This definition of Bose–Einstein condensation has the advantage that it is well defined for interacting systems. While Penrose and Onsager made explicit use of a thermodynamic limit procedure in their work, this definition can equally well be applied to a system of a finite number of particles. In the special case that $n_1^{(1)} = N$, the system is said to be *fully condensed* since then all particles occupy the same orbital. An N-boson state with $n_1^{(1)} = N$ is maximally coherent and satisfies Eq. 2.32, i.e. $|g^{(1)}| = 1$ and $g^{(2)} = 1 - 1/N$. States in which the largest natural orbital occupation number is very close to N are known as *depleted condensates*. To be definite we will call systems that satisfy $n_1^{(1)}/N > 95\%$ depleted condensates in this work. In case that more than one natural orbital occupation number is of the order of the number of particles,

$$n_1^{(1)}, n_2^{(1)}, \ldots = \mathcal{O}(N), \tag{2.41}$$

the condensate is said to be twofold *fragmented*, threefold *fragmented*, etc., according to Nozières and Saint James [18–19]. The sum over all $n_i^{(1)}$ for $i \geq 2$ is known as the *fragmentation* of the condensate. Fragmented condensates were initially thought to be unphysical, but it turns out that already the ground state of trapped Bose–Einstein condensates is often fragmented [7, 11, 18–27].

2.11 Classification of Interacting Regimes of Trapped Bose-Gases

In this section we briefly review the regimes of interacting Bose-gases. We have chosen to follow the classification scheme of Ref. [28, 29] here, since the results collected there on the statics of trapped Bose–Einstein condensates are the most rigorous available in the literature to date. For our purposes it is sufficient to restrict the discussion here to the one-dimensional (1D) case. We therefore write $\mathbf{r} = x$ and assume here that the interparticle interaction potential in Eq. 2.2 is given by $W(x - x') = \lambda_0 \delta(x - x')$.

The mean density $\bar{n}$ (of a stationary state) is defined as [28, 29]

$$\bar{n} = \int \rho(x;0)^2/N. \tag{2.42}$$

For homogeneous systems on an interval of length L the line density is simply $\bar{n} = N/L$. The parameter

$$\gamma = \lambda_0/\bar{n} \tag{2.43}$$

is then known as the *Lieb-Liniger parameter* [30]. The parameter γ appeared first in the exact treatment of a homogeneous Bose gas on a ring, where the parameter range $0 \leq \gamma \leq 2$ is known as the weak coupling limit, since the ground state energy in the thermodynamic limit can then be well approximated by perturbation theory [30]. Here we will call γ the Lieb-Liniger parameter also in the case of inhomogeneous systems, since the definition of $\bar{n}$ in Eq. 2.42 is generally applicable. In Refs. [28, 29] the following classification of one-dimensional interacting regimes was made:

- the ideal gas regime: $\gamma \ll N^{-2}$
- the 1D Gross–Pitaevskii regime : $\gamma \approx N^{-2}$
- the 1D Thomas-Fermi regime: $N^{-2} \ll \gamma \ll 1$
- the Lieb-Liniger regime: $\gamma \approx 1$
- the Girardeau-Tonks regime: $\gamma \gg 1$.

This classification is motivated by the rigorous mathematical results for the ground states of trapped condensates assuming an asymptotically homogeneous trapping potential and usually the limit $N \to \infty$ at constant $N\lambda_0$. We note that a potential $V(x)$ is asymptotically homogeneous, if $V(ax) = a^s V(x)$ for some $s > 0$ asymptotically in the limit $x \to \infty$. The first three regimes belong to the limit of weak interactions, with the third one being a special case of the second. The last two are characterized by strong interactions and were termed the 'true' 1D regimes.

It is important to realize that the naming convention above stems from the rigorous mathematical results obtained in the limit of an infinite number of particles. For any finite system this naming convention may or may not be appropriate. In Chap. 5 we will investigate the ground state of $N = 1{,}000$ bosons in a trap with $\gamma = 5 \times 10^{-5}$ and show explicitly that Gross–Pitaevskii theory fails there, although according to the above scheme it should be valid. In Chaps. 6 and 7 we will show how the two most popular theories of the field, Gross–Pitaevskii theory, and the Bose-Hubbard model fail to describe the dynamics of a bosonic Josephson junction deep within the regime where they are expected to be valid. We will also show that exciting new many-body effects exist at the beginning of the 1D Thomas-Fermi regime. In view of these facts the naming convention in the classification scheme above can be misleading when a finite number of particles is considered.

References

1. F. Schwabl, Advanced Quantum Mechanics, 4th edn. (Springer, Berlin, 2008)
2. P. Kramer, M. Saraceno, Geometry of the Time-Dependent Variational Principle in Quantum Mechanics. (Springer, Berlin, 1981)
3. A.J. Coleman, V.I. Yukalov, Reduced Density Matrices: Coulson's Challenge. (Springer, Berlin, 2000)
4. D.A. Mazziotti ed., *Reduced-Density-Matrix Mechanics: with Application to Many-Electron Atoms and Molecules*. Advances in Chemical Physics, vol 134 (Wiley, New York, 2007)
5. C.N. Yang, Rev. Mod. Phys. **34**, 694 (1962)
6. F. Sasaki, Phys. Rev. **138**, B1338 (1965)
7. K. Sakmann, A.I. Streltsov, O.E. Alon, L.S. Cederbaum, Phys. Rev. A **78**, 023615 (2008)
8. R.J. Glauber, Phys. Rev. **130**, 2529 (1963)
9. M. Naraschewski, R.J. Glauber, Phys. Rev. A **59**, 4595 (1999)
10. U.M. Titulaer, R.J. Glauber, Phys. Rev. **140**, B676 (1965)
11. A.I. Streltsov, O.E. Alon, L.S. Cederbaum, Phys. Rev. A **73**, 063626 (2006)
12. M. Orszag, Quantum Optics. (Springer, Berlin, 2000)
13. I. Bloch, T.W. Hänsch, T. Esslinger, Nature **403**, 166 (2000)
14. L.S. Cederbaum, A.I. Streltsov, Y.B. Band, O.E. Alon, cond-mat/0607556.
15. H. Xiong, S. Liu, M. Zhan, New J. Phys. **8**, 245 (2006)
16. L.S. Cederbaum, A.I. Streltsov, Y.B. Band, O.E. Alon, Phys. Rev. Lett. **98**, 110405 (2007)
17. O. Penrose, L. Onsager, Phys. Rev. **104**, 576 (1956)
18. P. Nozières, D. Saint James, J. Phys. **43**, 1133 (1982)
19. P. Nozières, in *Bose–Einstein Condensation*, ed. by A. Griffin, D.W. Snoke, and S. Stringari (Cambridge University Press, Cambridge, 1996)
20. R.W. Spekkens, J.E. Sipe, Phys. Rev. A **59**, 3868 (1999)
21. L.S. Cederbaum, A.I. Streltsov, Phys. Lett. A **318**, 564 (2003)
22. A.I. Streltsov, L.S. Cederbaum, N. Moiseyev, Phys. Rev. A **70**, 053607 (2004)
23. O.E. Alon, L.S. Cederbaum, Phys. Rev. Lett. **95**, 140402 (2005)
24. O.E. Alon, A.I. Streltsov, L.S. Cederbaum, Phys. Rev. Lett. **95**, 030405 (2005)
25. O.E. Alon, A.I. Streltsov, L.S. Cederbaum, Phys. Lett. A **347**, 88 (2005)
26. E.J. Mueller, T.-L. Ho, M. Ueda, G. Baym, Phys. Rev. A **74**, 033612 (2006)
27. P. Bader, U.R. Fischer, Phys. Rev. Lett. **103**, 060402 (2009)
28. E.H. Lieb, R. Seiringer, J.P. Solovej, J. Yngvason, The Mathematics of the Bose Gas and its Condensation. (Birkhäuser, Verlag, 2000)
29. E.H. Lieb, R. Seiringer, Phys. Rev. Lett. **91**, 150401 (2003)
30. E.H. Lieb, W. Liniger, Phys. Rev. **130**, 1605 (1963)

Chapter 3
General Methods for the Quantum Dynamics of Identical Bosons

We now turn to the description of the numerical methods used in this work. In Sect. 3.1 we give a self-contained explanation of the multiconfigurational time-dependent Hartree for bosons (MCTDHB) method. The MCTDHB method can be considered as the systematic many-body generalization of Gross–Pitaevskii theory. In Sect. 3.2 it is shown how the celebrated Gross–Pitaevskii equation can be recovered from MCTDHB as a special case. The MCTDHB method was invented and first implemented in Heidelberg. Please visit the http://MCTDHB.uni-hd.de. An Open-Source implementation of the MCTDHB algorithm can be found at http://OpenMCTDHB.uni-hd.de.

3.1 Multiconfigurational Time-Dependent Hartree for Bosons

3.1.1 Introduction

The dynamics of non-relativistic many-body quantum systems is generally determined by the time-dependent many-particle Schrödinger equation [1, 2]. Although this equation is linear, it can only be solved analytically in rare exceptions. Thus approximations and/or numerical methods are indispensable. A straightforward approach is to diagonalize the Hamiltonian in some time-independent basis set and to construct the solution of the Schrödinger equation at some time from the eigenvectors and energy eigenvalues thereby obtained. This approach which is known as *direct diagonalization* or *configuration interaction* is unfortunately limited to systems of small size and weak interaction strength. Moreover, the quality of the results obtained depends crucially on the chosen basis set. This problem has been known for a long time and in molecular physics a cure to this problem was proposed a long time ago. The idea is to use a time-adaptive optimized basis set [3–5]. A particularly efficient variant of this idea led to the multi-configuration time-dependent Hartree approach (MCTDH) which has been successfully applied to multi-dimensional dynamical systems consisting of distinguishable particles [6–12]. Of course, also

K. Sakmann, *Many-Body Schrödinger Dynamics of Bose–Einstein Condensates*, Springer Theses, DOI: 10.1007/978-3-642-22866-7_3,

systems of (few) indistinguishable particles can be investigated using MCTDH [13–20]. However, in order to treat systems consisting of a large number of identical particles it is crucial to exploit the symmetry of the many-body wave function under particle exchange.

The challenge to exploit this symmetry was first taken on for identical fermions. MCTDHF—the fermionic version of MCTDH—was developed independently by different groups several years ago [21–23]. MCTDHF is presently employed to study correlations effects in few-electron systems [24–26]. More recently, a bosonic version of MCTDH, called MCTDHB [27, 28] has been developed. The bosonic case is particularly interesting because—unlike fermions—a very large number of bosons can reside in a relatively small number of orbitals. Thereby it has become possible to investigate the true many-body dynamics of a large number of bosons quantitatively. As a first application of MCTDHB, the many-body dynamics of the splitting a condensate in a trap was studied in [27]. The MCTDHB method was then applied to problems such as the dynamics of condensates in double-wells [28], the investigation of correlations and coherence of trapped condensates [29], the dynamic formation of fragmentons and cat states [30, 31], the buildup of coherence between two initially independent subsystems [32], as well as the optimal control of number squeezing and atom interferometry [33–35]. Very recently the first numerically *exact* results on the quantum dynamics of a bosonic Josephson junction [36, 37] and for a quantum quench scenario [38] could be obtained, see Chaps. 6, 7 and 8. We note that it is possible to unite the fermionic and the bosonic methods in a common framework [39] and to extend the theory to mixtures of identical particles with particle conversion [40].

3.1.2 The MCTDHB Wave Function

The *ansatz* for the MCTDHB wave function $|\Psi(t)\rangle$ is of the most general form given in (2.15)

$$|\Psi(t)\rangle = \sum_{\vec{n}} C_{\vec{n}}(t)\, |n_1, n_2, \ldots, n_M; t\rangle\,, \tag{3.1}$$

where the summation in (3.1) runs over all $\binom{N+M-1}{N}$ *time-dependent* permanents, generated by distributing N bosons over M orbitals as defined in (2.14). In MCTDHB the orbitals $\{\phi_k(\mathbf{r}, t)\} = \{\phi_1(r, t), \ldots, \phi_M(\mathbf{r}, t)\}$ are determined variationally which makes them time-dependent, see below.

The finite size representation, (2.16), of the field operator, (2.3), then takes on the form

$$\hat{\boldsymbol{\Psi}}_M(\mathbf{r}; t) = \sum_{k=1}^{M} \hat{b}_k(t)\phi_k(\mathbf{r}, t), \tag{3.2}$$

which is also time-dependent.

3.1.3 Derivation of the MCTDHB Equations

To derive the set of equations-of-motion for the multi-configurational time-evolution of identical bosons [27, 28, 39], we employ the Lagrangian formulation of the time-dependent variational principle [41]. Thus, we substitute the many-body *ansatz* (3.1) into (2.12) and enforce the constraint of orthonormality between the orbitals $\{\phi_k(\mathbf{r},t)\}$ by Lagrange multipliers $\mu_{kj}(t)$. The functional action of the time-dependent Schrödinger equation then reads:

$$S\left[\{C_{\vec{n}}\},\{\phi_k\}\right]=\int dt\left\{\left\langle\Psi\left|\hat{H}-i\frac{\partial}{\partial t}\right|\Psi\right\rangle-\sum_{k,j}^{M}\mu_{kj}(t)\left[\langle\phi_k|\phi_j\rangle-\delta_{kj}\right]\right\}. \quad (3.3)$$

We recall that the orbitals $\{\phi_k(\mathbf{r},t)\}$ and coefficients $\{C_{\vec{n}}(t)\}$ are independent variables (arguments) of the action (3.3). Our strategy is first to take expectation values and only subsequently perform the variation and require stationarity of the action with respect to the arguments $\{\phi_k(\mathbf{r},t)\}$ and $\{C_{\vec{n}}(t)\}$.

To perform the variation of the action (3.3) with respect to the orbitals we express the expectation value of $\hat{H}-i\frac{\partial}{\partial t}$ appearing in (3.3) in a form which explicitly depends on the orbitals. This can be done by resorting to the first and second order RDMs $\rho^{(1)}(\mathbf{r}_1|\mathbf{r}_1';t)$ and $\rho^{(2)}(\mathbf{r}_1,\mathbf{r}_2|\mathbf{r}_1',\mathbf{r}_2';t)$, (2.33) and (2.34). The expectation value in (3.3) then reads

$$\left\langle\Psi\left|\hat{H}-i\frac{\partial}{\partial t}\right|\Psi\right\rangle=\sum_{k,q=1}^{M}\rho_{kq}\left[h_{kq}-\left(i\frac{\partial}{\partial t}\right)_{kq}\right]+\frac{1}{2}\sum_{k,s,l,q=1}^{M}\rho_{kslq}W_{ksql}-i\sum_{\vec{n}}C_{\vec{n}}^{*}\frac{\partial C_{\vec{n}}}{\partial t}, \quad (3.4)$$

where the time-derivative $i\frac{\partial}{\partial t}$ is written as a one-body operator,

$$i\frac{\partial}{\partial t}=\sum_{k,q}\hat{b}_k^\dagger\hat{b}_q\left(i\frac{\partial}{\partial t}\right)_{kq}, \quad \left(i\frac{\partial}{\partial t}\right)_{kq}=i\int\phi_k^*(\mathbf{r},t)\frac{\partial\phi_q(\mathbf{r},t)}{\partial t}d\mathbf{r} \quad (3.5)$$

and for brevity we have dropped the time-argument everywhere. The matrix elements h_{kq} and W_{ksql} were defined in (2.10).

Representation (3.4) is very useful because the only explicit dependence on the orbitals $\{\phi_k(\mathbf{r},t)\}$ is contained in the matrix elements h_{kq}, $\left(i\frac{\partial}{\partial t}\right)_{kq}$ and W_{ksql}, whereas the elements ρ_{kq} and ρ_{kslq} of the first and second order RDMs do not depend explicitly on the orbitals. See Appendices C and D for explicit expressions of these quantities. It is convenient to collect the elements ρ_{kq} in a matrix $\boldsymbol{\rho}(t)=\left\{\rho_{kq}(t)\right\}$.

The variation of the functional action (3.3) with respect to the orbitals is now straightforward. Using the fact that the $\{\phi_k(\mathbf{r},t)\}$ are orthonormal functions to eliminate the Lagrange multipliers $\mu_{kj}(t)$, we obtain the following set of equations-of-motion for the time-dependent orbitals in which the particles reside, $j=1,\ldots,M$:

$$\hat{\mathbf{P}}i\left|\dot{\phi}_j\right\rangle = \hat{\mathbf{P}}\left[\hat{h}\left|\phi_j\right\rangle + \sum_{k,s,l,q=1}^{M}\{\boldsymbol{\rho}(t)\}^{-1}_{jk}\,\rho_{kslq}\hat{W}_{sl}\left|\phi_q\right\rangle\right],$$

$$\hat{\mathbf{P}} = 1 - \sum_{j'=1}^{M}\left|\phi_{j'}\right\rangle\left\langle\phi_{j'}\right|, \tag{3.6}$$

where

$$\hat{W}_{sl}(\mathbf{r},t) = \int \phi_s^*(\mathbf{r}',t)\hat{W}(\mathbf{r}-\mathbf{r}')\phi_l(\mathbf{r}',t)d\mathbf{r}' \tag{3.7}$$

are *local* time-dependent potentials and $\dot{\phi}_j \equiv \frac{\partial\phi_j}{\partial t}$. Examining (3.6) we see that eliminating the Lagrange multipliers $\mu_{kj}(t)$ has emerged as a projection operator $\hat{\mathbf{P}}$ onto the subspace orthogonal to that spanned by the orbitals. This projection appears both on the left- and right-hand-sides of (3.6), making (3.6) a cumbersome coupled system of integro-differential non-linear equations.

To simplify the equations-of-motion (3.6) we recall that the many-body wave function (2.15) is invariant to unitary transformations of the orbitals, compensated by 'reverse' transformations of the coefficients. Fortunately, there exists *one* specific unitary transformation which guarantees without introducing further constraints that [6, 7]

$$\left\langle\phi_k|\dot{\phi}_q\right\rangle = 0, \quad k,q = 1,\ldots,M \tag{3.8}$$

are satisfied at any time. Obviously, if conditions (3.8) are satisfied at any time, the orbitals remain orthonormal functions at any time. This representation simplifies considerably the equations-of-motion (3.6) for the orbitals, $j = 1,\ldots,M$:

$$\begin{aligned} i\left|\dot{\phi}_j\right\rangle &= \hat{\mathbf{P}}\left[\hat{h}\left|\phi_j\right\rangle + \sum_{k,s,l,q=1}^{M}\{\boldsymbol{\rho}(t)\}^{-1}_{jk}\,\rho_{kslq}\hat{W}_{sl}\left|\phi_q\right\rangle\right], \\ \hat{\mathbf{P}} &= 1 - \sum_{j'=1}^{M}\left|\phi_{j'}\right\rangle\left\langle\phi_{j'}\right|. \end{aligned} \tag{3.9}$$

The projector $\hat{\mathbf{P}}$ remaining on the right-hand-side of (3.9) makes it clear that conditions (3.8) are indeed met at any time throughout the propagation of the orbitals. In practice, the meaning of these conditions is that the temporal changes of the $\{\phi_k(\mathbf{r},t)\}$ are always orthogonal to the $\{\phi_k(\mathbf{r},t)\}$ themselves. This property introduced by the MCTDH developers [6, 7] generally makes the time propagation of (3.9) robust and stable and can thus be exploited to maintain accurate propagation results at lower computational costs.

To complete the derivation, we perform the variation of (3.3) with respect to the coefficients which is easily done after expressing the expectation value of $\hat{H} - i\frac{\partial}{\partial t}$ in a form which explicitly depends on the $\{C_{\vec{n}}(t)\}$. The following result then emerges:

$$\mathcal{H}(t)\mathbf{C}(t) = i\frac{\partial \mathbf{C}(t)}{\partial t} \tag{3.10}$$

with

$$\mathcal{H}_{\vec{n}\vec{n}'}(t) = \left\langle n_1, n_2, \ldots, n_M; t \left| \hat{H} - i\frac{\partial}{\partial t} \right| n'_1, n'_2, \ldots, n'_M; t \right\rangle, \tag{3.11}$$

where the vector $\mathbf{C}(t)$ collects the coefficients $\{C_{\vec{n}}(t)\}$. The matrix elements of $\hat{H} - i\frac{\partial}{\partial t}$ with respect to two general configurations $|n_1, n_2, \ldots, n_M; t\rangle$ and $|n'_1, n'_2, \ldots, n'_M; t\rangle$ are easily evaluated using the general rules for permanent expectation values [42]. The rules used are collected in Appendix D. Finally, making use of conditions (3.8) we obtain the familiar equations-of-motion for the propagation of the coefficients,

$$\mathbf{H}(t)\mathbf{C}(t) = i\frac{\partial \mathbf{C}(t)}{\partial t} \tag{3.12}$$

with

$$H_{\vec{n}\vec{n}'}(t) = \left\langle n_1, n_2, \ldots, n_M; t \left| \hat{H} \right| n'_1, n'_2, \ldots, n'_M; t \right\rangle. \tag{3.13}$$

The coupled equation sets (3.6) for the orbitals $\{\phi_j(\mathbf{r}, t)\}$ and (3.10) for the expansion coefficients $\{C_{\vec{n}}(t)\}$, or equivalently, (3.9) and (3.12) constitute the MCTDHB equations [27, 28]. For an implementation the set of coupled equations (3.9) and (3.12) are better suited. Very recently a parallel version of MCTDHB has been implemented using a novel mapping technique [43]. We note that by propagation in imaginary time the MCTDHB equations also allow to determine ground and excited states of interacting many-boson systems.

The MCTDHB equations-of-motion become an exact representation of the time-dependent many-particle Schrödinger equation in the limit where M goes to infinity. In practice, M is of course finite, This is where the employment of time-dependent, variationally determined orbitals—which is at the heart of MCTDHB—is of great advantage. As mentioned earlier the number of permanents used in MCTDHB is $\binom{N+M-1}{N}$. This is a very rapidly growing function of of the number of orbitals and the number of bosons, and actually nothing specific to MCTDHB. It is simply the number of permanents that can be assembled from M orbitals and N particles. However, any other method using time-independent orbitals will need substantially more orbitals, i.e. a larger M, to achieve the same accuracy as MCTDHB. In Appendix F we have included a comparison between an expansion in time-independent and optimized, time-dependent orbitals which illustrates this point.

MCTDHB therefore provides a mathematically sound and optimal procedure to converge to the exact solution of the many-body Schrödinger equation. This is done as follows. A computation for N bosons is first carried out using just one orbital $M=1$. As we will show in Sect. 3.2 the restriction of MCTDHB to $M=1$ orbital results in the Gross–Pitaevskii equation. The same computation is then repeated using increasingly more orbitals until no appreciable change in the results is detected, i.e. until the results have converged. If convergence is obtained the results are exact.

3.2 Gross–Pitaevskii Theory

In this section we derive the celebrated Gross–Pitaevskii mean-field theory and show how it can be obtained as a special case of MCTDHB. We follow the exposition of Ref. [44]. In Gross–Pitaevskii theory the *ansatz* for the many-boson wave function $|\Psi(t)\rangle$ is

$$|\Psi(t)\rangle = |N; t\rangle\,, \tag{3.14}$$

By comparison with (2.15) we find that the Gross–Pitaevskii ansatz amounts to choosing $M=1$, i.e. the many-body wave function is approximated a single *time-dependent* permanent in which all N particles occupy the same orbital. There are no coefficients $C_{\vec{n}}(t)$ to be determined in this ansatz, because only a single permanent is used.

The finite size representation, (2.16), of the field operator, (2.3), then takes on the form

$$\hat{\mathbf{\Psi}}_M(\mathbf{r}; t) = \hat{b}_1(t)\phi_1(\mathbf{r}, t), \tag{3.15}$$

i.e. the field operator is approximated by an expansion over precisely one time-dependent orbital.

As an interparticle interaction potential we choose $W(\mathbf{r} - \mathbf{r}') = \lambda_0\delta(\mathbf{r} - \mathbf{r}')$. In order to derive equations of motion we substitute the Gross–Pitaevskii ansatz, (3.14), into the action functional of the time-dependent many-body Schrödinger equation, (3.3), and perform a variation with respect to the orbital. We thereby obtain

$$N\Big[h(\mathbf{r}) + \lambda_0(N-1)|\phi_1(\mathbf{r}, t)|^2\Big]\phi_1(\mathbf{r}, t) = \mu_{11}(t)\phi_1(\mathbf{r}, t) + iN\dot{\phi}_1(\mathbf{r}, t). \tag{3.16}$$

The Lagrange multiplier $\mu_{11}(t)$ appearing on the right-hand side of (3.16) can be absorbed into ϕ_1 by shifting its global phase

$$\phi_1(\mathbf{r}, t) = \tilde{\phi}_1(\mathbf{r}, t)e^{-i\int^t dt'\mu_{11}(t')/N}. \tag{3.17}$$

Substituting (3.17) back into (3.16) and dropping the tilde and the subscript on $\tilde{\phi}_1(x, t)$ we arrive at the celebrated Gross–Pitaevskii equation [45–47] as it is standardly written

$$i\dot{\phi}(\mathbf{r}, t) = \Big[h(\mathbf{r}) + \lambda_0(N-1)|\phi(\mathbf{r}, t)|^2\Big]\phi(\mathbf{r}, t). \tag{3.18}$$

Instead of eliminating $\mu_{11}(t)$ by making the phase choice in (3.17) one can of course first determine $\mu_{11}(t)$ by multiplying (3.16) with $\int \phi_1^*(\mathbf{r}, t)$ from the left. Substituting the result for $\mu_{11}(t)$ back into (3.16) yields the MCTDHB orbital equation of motion for $M=1$ in the form given in (3.6)

$$\hat{\mathbf{P}}i\dot{\phi}_1(\mathbf{r}, t) = \hat{\mathbf{P}}\Big[h(\mathbf{r}) + \lambda_0(N-1)|\phi_1(\mathbf{r}, t)|^2\Big]\phi_1(\mathbf{r}, t). \tag{3.19}$$

By shifting the global phase of the orbital through the assignment

$$\phi_1(\mathbf{r}, t) = e^{+\int^t dt' <\phi_1|\dot{\phi}_1>} \tilde{\phi}_1(\mathbf{r}, t) \tag{3.20}$$

we arrive after substituting (3.20) into (3.19) and dropping the tilde and the subscript on $\tilde{\phi}_1(x, t)$ at

$$i\dot{\phi}(\mathbf{r}, t) = \hat{\mathbf{P}}\Big[h(\mathbf{r}) + \lambda_0(N-1)|\phi(\mathbf{r}, t)|^2\Big]\phi(\mathbf{r}, t). \tag{3.21}$$

Obviously, (3.21) and (3.18) are equivalent as they only differ in the assignment of a global time-dependent phase for the orbital. Equation (3.21) is nothing else but the MCTDHB orbital equation of motion for $M=1$ in the form given in (3.9).

We have thereby shown that the Gross–Pitaevskii equation in its conventional form, (3.18) is equivalent to MCTDHB with $M=1$. For the numerical solution of the Gross–Pitaevskii equation the form (3.21) is clearly preferable as the projector $\hat{\mathbf{P}}$ ensures that the change in the orbital at every time step is orthogonal to the orbital itself. If (3.21) is implemented no computing time is wasted on a meaningless integration of the phase of $\phi(\mathbf{r}, t)$ which usually changes rapidly if (3.18) is solved directly.

It is clear that the Gross–Pitaevskii ansatz for the wave function, (3.14),is only the simplest possible ansatz for the many-body wave function in (2.15). More than one time-dependent orbital will generally be necessary in order to obtain qualitatively correct results, see the discussion at the end of Sect. 3.1. It is also instructive to take a look at the finite size representation of the Gross–Pitaevskii field operator, (3.15). If Gross–Pitaevskii theory is valid, only one single term, $\hat{b}_1(t)\phi_1(\mathbf{r}, t)$, is relevant in the exact expansion of the field operator, (2.3), at any time. For all practical purposes it is then irrelevant whether the exact expression, (2.3), or (3.15) is used in the computation of any physical quantity. Generally, this is of course not the case.

For a given system of identical bosons it is possible to assess *a posteriori* whether Gross–Pitaevskii theory is valid by solving the same problem using MCTDHB with $M>1$. If the results using $M>1$ orbitals do not differ noticeably from the Gross–Pitaevskii results, Gross–Pitaevskii theory is valid.

It would be of great interest to have a rigorous criterion that allows to determine *a priori* whether or not Gross–Pitaevskii theory is applicable for a given system of a finite number of bosons. No such criterion is known to date, not even for the ground state. In the limit of an infinite number of particles $N \to \infty$ and keeping $N\lambda_0$ constant such criteria do exist, see the classification scheme in (Sect. 2.11) and Refs. [48, 49]. In the same limit it can even be proven that Gross–Pitaevskii theory is valid for the dynamics of a condensate released from a trap [50]. For a finite number of interacting bosons such statements are generally incorrect, as can already be concluded from the analysis of exactly solvable models [51–54]. In inhomogeneous systems already the ground state of a finite number of particles is often fragmented even if the interaction strength is very small [29, 42, 55–60].

References

1. *Time-Dependent Methods for Quantum Dynamics*, ed. by K. C. Kulander (Elsevier, Amsterdam, 1991)
2. *Many-Particle Quantum Dynamics in Atomic and Molecular Fragmentation*, ed. by J. Ullrich, V. P. Shevelko (Springer, Berlin, 2003)
3. N. Makri, W.H. Miller, J. Chem. Phys. **87**, 5781 (1987)
4. R. Kosloff, A.D. Hammerich, M.A. Ratner, in *Large Finite Systems: Proceedings of the Twentieth Jerusalem Symposium of Quantum Chemistry and Biochemistry*, ed. by J. Jortner, B. Pullman (Reidel, Dordrecht, 1987)
5. A.D. Hammerich, R. Kosloff, M.A. Ratner, Chem. Phys. Lett. **171**, 97 (1990)
6. H.-D. Meyer, U. Manthe, L.S. Cederbaum, Chem. Phys. Lett. **165**, 73 (1990)
7. U. Manthe, H.-D. Meyer, L.S. Cederbaum, J. Chem. Phys. **97**, 3199 (1992)
8. G.A. Worth, H.-D. Meyer, L.S. Cederbaum, J. Chem. Phys. **109**, 3518 (1998)
9. M.H. Beck, A. Jäckle, G.A. Worth, H.-D. Meyer, Phys. Rep. **324**, 1 (2000)
10. van R. Harrevelt, U. Manthe, J. Chem. Phys. **123**, 064106 (2005)
11. H.-D. Meyer, G.A. Worth, Theor. Chem. Acc. **109**, 251 (2003)
12. *Multidimensional Quantum Dynamics: MCTDH Theory and Applications*, ed. by H.-D. Meyer, F. Gatti, G. A. Worth (WILEY-VCH, Weinheim, 2009)
13. S. Zöllner, H.-D. Meyer, P. Schmelcher, Phys. Rev. A **74**, 053612 (2006)
14. S. Zöllner, H.-D. Meyer, P. Schmelcher, Phys. Rev. A **74**, 063611 (2006)
15. S. Zöllner, H.-D. Meyer, P. Schmelcher, Phys. Rev. A **75**, 043608 (2007)
16. C. Matthies, S. Zöllner, H.-D. Meyer, P. Schmelcher, Phys. Rev. A **76**, 023602 (2007)
17. S. Zöllner, H.-D. Meyer, P. Schmelcher, Phys. Rev. Lett. **100**, 040401 (2008)
18. S. Zöllner, H.-D. Meyer, P. Schmelcher, Phys. Rev. A **78**, 013621 (2008)
19. S. Zöllner, H.-D. Meyer, P. Schmelcher, Phys. Rev. A **78**, 013629 (2008)
20. A.U.J. Lode, A.I. Streltsov, O.E. Alon, H.-D. Meyer, L.S. Cederbaum, J. Phys. B **42**, 044018 (2009)
21. J. Zanghellini, M. Kitzler, C. Fabian, T. Brabec, A. Scrinzi, Laser Phys. **13**, 1064 (2003)
22. T. Kato, H. Kono, Chem. Phys. Lett. **392**, 533 (2004)
23. M. Nest, T. Klamroth, P. Saalfrank, J. Chem. Phys. **122**, 124102 (2005)
24. J. Caillat, J. Zanghellini, M. Kitzler, O. Koch, W. Kreuzer, A. Scrinzi, Phys. Rev. A **71**, 012712 (2005)
25. M. Kitzler, J. Zanghellini, Ch. Jungreuthmayer, M. Smits, A. Scrinzi, T. Brabec, Phys. Rev. A **70**, 041401 (2004)
26. M. Nest, Phys. Rev. A **73**, 023613 (2006)
27. A.I. Streltsov, O.E. Alon, L.S. Cederbaum, Phys. Rev. Lett. **99**, 030402 (2007)
28. O.E. Alon, A.I. Streltsov, L.S. Cederbaum, Phys. Rev. A **77**, 033613 (2008)
29. K. Sakmann, A.I. Streltsov, O.E. Alon, L.S. Cederbaum, Phys. Rev. A **78**, 023615 (2008)
30. A.I. Streltsov, O.E. Alon, L.S. Cederbaum, Phys. Rev. Lett. **100**, 130401 (2008)
31. A.I. Streltsov, O.E. Alon, L.S. Cederbaum, Phys. Rev. A **80**, 043616 (2009)
32. O.E. Alon, A.I. Streltsov, L.S. Cederbaum, Phys. Lett. A **373**, 301 (2009)
33. J. Grond, J. Schmiedmayer, U. Hohenester, Phys. Rev. A **79**, 021603(R) (2009)
34. J. Grond, von G. Winckel, J. Schmiedmayer, U. Hohenester, Phys. Rev. A **80**, 053625 (2009)
35. J. Grond, J. Schmiedmayer, U. Hohenester, New J. Phys. **12**, 065036 (2010)
36. K. Sakmann, A.I. Streltsov, O.E. Alon, L.S. Cederbaum, Phys. Rev. Lett. **103**, 220601 (2009)
37. K. Sakmann, A.I. Streltsov, O.E. Alon, L.S. Cederbaum, Phys. Rev A **82**, 013620 (2010)
38. K. Sakmann, A.I. Streltsov, O.E. Alon, L.S. Cederbaum, New J. Phys. **13**, 043003 (2011)
39. O.E. Alon, A.I. Streltsov, L.S. Cederbaum, J. Chem. Phys. **127**, 154103 (2007)
40. O.E. Alon, A.I. Streltsov, L.S. Cederbaum, Phys. Rev. A **79**, 022503 (2009)
41. *Geometry of the Time-Dependent Variational Principle in Quantum Mechanics*, ed. by P. Kramer, M. Saraceno (Springer, Berlin, 1981)
42. A.I. Streltsov, O.E. Alon, L.S. Cederbaum, Phys. Rev. A **73**, 063626 (2006)

43. A.I. Streltsov, O.E. Alon, L.S. Cederbaum, Phys. Rev. A **81**, 022124 (2010)
44. O.E. Alon, A.I. Streltsov, L.S. Cederbaum, Phys. Lett. A **362**, 453 (2007)
45. E.P. Gross, Nuovo Cimento **20**, 454 (1961)
46. L.P. Pitaevskii, Zh. Eksp. Teor. Fiz. **40**, 646 (1961)
47. L.P. Pitaevskii, Sov. Phys. JETP **13**, 451 (1961)
48. *The Mathematics of the Bose Gas and its Condensation*, ed. by E. H. Lieb, R. Seiringer, J. P. Solovej, J. Yngvason, (Birkhäuser Verlag, Basel, 2000)
49. E.H. Lieb, R. Seiringer, Phys. Rev. Lett. **91**, 150401 (2003)
50. L. Erdős, B. Schlein, H. Yau, Phys. Rev. Lett. **98**, 040404 (2007)
51. M. Girardeau, J. Math. Phys. **1**, 516 (1960)
52. E.H. Lieb, W. Liniger, Phys. Rev. **130**, 1605 (1963)
53. J.B. McGuire, J. Math. Phys. **5**, 622 (1964)
54. K. Sakmann, A.I. Streltsov, O.E. Alon, L.S. Cederbaum, Phys. Rev. A **72**, 033613 (2005)
55. R.W. Spekkens, J.E. Sipe, Phys. Rev. A **59**, 3868 (1999)
56. L.S. Cederbaum, A.I. Streltsov, Phys. Lett. A **318**, 564 (2003)
57. A.I. Streltsov, L.S. Cederbaum, N. Moiseyev, Phys. Rev. A **70**, 053607 (2004)
58. O.E. Alon, L.S. Cederbaum, Phys. Rev. Lett. **95**, 140402 (2005)
59. O.E. Alon, A.I. Streltsov, L.S. Cederbaum, Phys. Lett. A **347**, 88 (2005)
60. E.J. Mueller, T.-L. Ho, M. Ueda, G. Baym, Phys. Rev. A **74**, 033612 (2006)

Chapter 4
Lattice Models for the Quantum Dynamics of Identical Bosons

In the previous Chapter two general theories, MCTDHB and Gross–Pitaevskii theory were discussed, and it was shown that Gross–Pitaevskii theory is a special case of MCTDHB. These methods are general in the sense that no assumptions were made about the system of bosons apart from the *ansatz* for the many-body wave function. In this Chapter, we discuss models that are popular in the description of ultracold bosons in (quasi-) periodic potentials, namely the Bose–Hubbard model and the discrete nonlinear Schrödinger equation. These two methods have in common that both employ Wannier functions as a single-particle basis. For our purposes it will be sufficient to restrict the discussion to the case of 1D-double-well potentials. We begin with a discussion of the Wannier functions and the parameters that appear in the two models.

4.1 Wannier Functions of a Double-Well Potential

Given a strictly periodic potential in 1D, the single-particle eigenfunctions of the potential are known as Bloch waves. The so called *Wannier functions* are then constructed as linear superpositions of the Bloch waves. Wannier functions have the appealing property to be real and localized at the minima of the periodic potential [1]. It is customary to apply the concept of Wannier functions also to not strictly-periodic potentials, such as optical lattices, double- and multi-well potentials. In the context of bosonic Josephson junctions—the topic of Chaps. 6 and 7—double-well potentials are of particular importance. We therefore consider a symmetric double-well potential

$$V(-x) = V(x) \tag{4.1}$$

and denote the eigenfunctions by $\phi_1, \phi_2, \phi_3, \ldots$ with energies $e_1, e_2, e_3, \ldots$. The single-particle ground state of $V(x)$ is symmetric $\phi_1(-x) = \phi_1(x)$ and the first excited state antisymmetric $\phi_2(-x) = -\phi_2(x)$. Left- and right-localized orbitals

K. Sakmann, *Many-Body Schrödinger Dynamics of Bose–Einstein Condensates*,
Springer Theses, DOI: 10.1007/978-3-642-22866-7_4,

$$\phi_{L,R}(x) = \frac{1}{\sqrt{2}}\Big(\phi_1(x) \pm \phi_2(x)\Big) \tag{4.2}$$

can then be constructed from $\phi_1(x)$ and $\phi_2(x)$. The orbitals $\phi_{L,R}$ are orthonormal to one another and satisfy $\phi_L(-x) = \phi_R(x)$. For sufficiently high barriers the orbitals ϕ_L and ϕ_R are localized in the left and the right well of the double-well potential. Using the orbitals $\phi_{L,R}$ as a single-particle basis allows for an intuitive picture of bosons populating either the left or the right orbital.

From the orbitals $\phi_{L,R}$, the one-body Hamiltonian $h(x) = -\frac{1}{2}\frac{\partial^2}{\partial x^2} + V_{dw}(x)$, and the interaction potential $W(x - x') = \lambda_0 \delta(x - x')$ the matrix elements

$$\begin{aligned} U &= W_{RRRR} = W_{LLLL} = \lambda_0 \int dx |\phi_L(x)|^4 \\ J &= -h_{RL} = -h_{LR} = \frac{e_2 - e_1}{2} = -\int dx \phi_L^*(x) h(x) \phi_R(x) \\ \epsilon &= h_{LL} = h_{RR} = \frac{e_1 + e_2}{2} = \int dx \phi_L^*(x) h(x) \phi_L(x) \end{aligned} \tag{4.3}$$

can be computed that will be useful in following sections. U parameterizes the on-site interaction energy, J the hopping from site to site and ϵ is the energy of a single particle localized residing in either of the orbitals ϕ_L *or* ϕ_R. In a lattice ϵ lies in the middle of the lowest band which consists here of just two states and has a width of $2J = e_2 - e_1$. The energy difference between the first and the second excited state

$$e_{gap} = e_3 - e_2 \tag{4.4}$$

is the equivalent of the first band gap in a lattice.

4.2 The Two-Mode Gross–Pitaevskii Model

In the literature on bosonic Josephson junctions the Gross–Pitaevskii mean-field approximation plays a crucial role since equations of motion that resemble the superconducting Josephson junction equations can be derived from it, if a few additional assumptions are made [2–5]. The procedure described here, can be easily be applied to lattices and the resulting equations are also known as discrete nonlinear Schrödinger equations. We briefly review the procedure and start from the Gross–Pitaevskii equation, Eq. 3.18. Let $\phi(x,t)$ be the Gross–Pitaevskii orbital in Eq. 3.18 and $V(x)$ a symmetric double-well potential, as in Eq. 4.1. In the two-mode Gross–Pitaevskii model the *ansatz*

$$\phi(x,t) = d_L(t)\phi_L(x) + d_R(t)\phi_R(x) \tag{4.5}$$

is then made. The finite size representation (3.15) of the Gross–Pitaevskii field operator then reads

$$\hat{\boldsymbol{\Psi}}_M(x;t) = \hat{b}_1(t)\Big(d_L(t)\phi_L(x) + d_R(t)\phi_R(x)\Big). \tag{4.6}$$

Substituting the *ansatz* (4.5) into Eq. 3.18 multiplying with $\int \phi^*_{L,R}$ from the left, keeping only the lowest order terms only and approximating $N - 1 \approx N$ results in the two-mode Gross–Pitaevskii model [2, 3]

$$\begin{aligned} i\frac{d}{dt}d_L(t) &= UN|d_L(t)|^2 d_L(t) - Jd_R(t) \\ i\frac{d}{dt}d_R(t) &= UN|d_R(t)|^2 d_R(t) - Jd_L(t). \end{aligned} \tag{4.7}$$

The parameters U and J defined in Eq. 4.3 and we have chosen the origin of the energy such that $\epsilon = 0$. In a different context an exact analytical solution of the Eq. 4.7 was found a long time ago [6]. This two-mode model can be mapped onto a classical nonlinear pendulum, see Refs.[3, 4]. The parameter

$$\Lambda = \frac{UN}{2J} \tag{4.8}$$

characterizes the interaction strength in the two-mode Gross–Pitaevskii model.

The analytic solution of the two-mode Gross–Pitaevskii model predicts that from some critical interaction strength Λ_c onwards the density remains trapped in one well forever [2, 3, 6]. More precisely, for $\Lambda > \Lambda_c$ the long term average of the density in each of the wells can be different from $N/2$:

$$\lim_{T\to\infty} \frac{1}{T}\int_0^T dtN|d_L(t)|^2 \neq \frac{N}{2}, \tag{4.9}$$

at least within the two-mode Gross–Pitaevskii model. This phenomenon is known as *self-trapping*. Self-trapping as it is defined above is a unique feature of the two-mode Gross–Pitaevskii model. In this strict form self-trapping does not exist, because the trapping potential is symmetric and the many-body eigenstates are parity eigenstates. We will come back to this point at a later stage. Nevertheless, the time scale on which self-trapping is lost can become very large in the full many-body description.

4.3 The Bose–Hubbard Model

In this section, we briefly review the Bose–Hubbard model [2, 7–9], which is the standard many-body model for the description of ultracold bosons in double-well, multi-well and (quasi-) periodic external potentials. We treat here only the case of a 1D double-well potential. More detailed accounts can be found, e.g. in Refs. [2, 5, 10, 11].

We consider again a symmetric double-well potential $V(x)$, as in Eq. 4.1. In the Bose–Hubbard model the *ansatz* for wave function $|\Psi(t)\rangle$ is

$$|\Psi(t)\rangle = \sum_{\vec{n}} C_{\vec{n}}(t)\,|n_L, n_R\rangle\,, \tag{4.10}$$

where $n_L + n_R = N$ is fixed and the summation in (4.10) runs over all $\binom{N+1}{N} = N+1$ time-independent permanents, generated by distributing N bosons over the two orbitals ϕ_L and ϕ_R:

$$|n_L, n_R\rangle = \frac{1}{\sqrt{n_L! n_R!}} \left(\hat{b}_L^\dagger\right)^{n_L} \left(\hat{b}_R^\dagger\right)^{n_R} |vac\rangle\,. \tag{4.11}$$

With the *ansatz* (4.10) the finite M representation of the field operator, Eq. 2.16 takes on the form

$$\hat{\boldsymbol{\Psi}}_M(x) = \hat{b}_L\phi_L(x) + \hat{b}_R\phi_R(x). \tag{4.12}$$

Note that Eqs. 4.10 and 4.11 are qualitatively different from those of the most general ansatz, given in Eqs. 2.14 and 2.15 because the orbitals here are not allowed to depend on time. This implies that significantly more bands are generally needed than when time-dependent orbitals are used. In Appendix F we have included an example which illustrates this point. Substituting (4.12) into Eq. 2.8 the many-body Hamiltonian takes on the form

$$\hat{H} = \sum_{k,q=L,R} \hat{b}_k^\dagger \hat{b}_q h_{kq} + \frac{1}{2} \sum_{k,s,l,q=L,R} \hat{b}_k^\dagger \hat{b}_s^\dagger \hat{b}_l \hat{b}_q W_{ksql}, \tag{4.13}$$

where now all matrix elements h_{kq} and W_{ksql} are time-independent and real in contrast to the general case, see Eq. 2.9. As in Sect. 3.2 we choose $W(x-x') = \lambda_0\delta(x-x')$ as an interparticle interaction potential. If the central barrier of the potential $V(x)$ is high enough, all two-body matrix elements in which not all indices are equal, e.g. W_{RLLL}, are much smaller than the matrix element $U = W_{LLLL} = W_{RRRR}$. Therefore, all two-body matrix elements except for W_{LLLL} and W_{RRRR} are neglected in the Bose–Hubbard model. With the choice $\epsilon = 0$ we finally arrive at the Bose–Hubbard Hamiltonian for a double well potential as it is standardly written

$$\hat{H}_{BH} = -J\left(\hat{b}_L^\dagger \hat{b}_R + b_R^\dagger \hat{b}_L\right) + \frac{U}{2}\left(\hat{b}_L^\dagger \hat{b}_L^\dagger \hat{b}_L \hat{b}_L + \hat{b}_R^\dagger \hat{b}_R^\dagger \hat{b}_R \hat{b}_R\right). \tag{4.14}$$

In the above derivation it was first assumed that the many-body wave function can be expanded at all times in the Wannier functions of the lowest band, see Eq. 4.10. Secondly, it was assumed that the Hamiltonian in Eq. 4.13 can be further simplified by neglecting all off-diagonal terms in the two-body part. Whether these approximations are justified for a given system can be determined *a posteriori* by comparison with the exact solution of the many-body Schrödinger equation for the same problem.

However, it would be highly desirable to have a criterion that allows to assess the applicability of the Bose–Hubbard model *a priori*. No rigorous criterion is known to date. The available criteria all make explicit use of the parameters U and J and thereby of the noninteracting system. For example, the standard criterion requires that [2]

$$\frac{NU}{\Delta E} \ll 1, \tag{4.15}$$

where ΔE is the single-particle level spacing of either of the wells in the harmonic oscillator approximation. The criterion (4.15) can be motivated by considering the many-boson states that can be constructed from the lowest band as a set of quasi-degenerate states and applying perturbation theory. The state the highest in energy has an energy Ne_2. The lowest state in the next band has an energy Ne_3, so the energy difference between these states is $N(e_3 - e_2) = Ne_{gap}$. By considering the operator $\sum_{i<j} \lambda_0 \delta(x_i - x_j)$ as a perturbation to the noninteracting system, the energy corrections to the unperturbed many-boson states of the lowest band are of the order N^2U in first order perturbation theory. A necessary requirement for the applicability of quasi-degenerate perturbation theory is that the first order energy correction is small compared to the distance to other energy levels, in formulas

$$\frac{NU}{e_{gap}} \ll 1. \tag{4.16}$$

The criterion (4.16) is stricter than (4.15) since e_{gap} is always smaller than ΔE. For a state with all N bosons at a given site the energy needed to add another particle at that site is NU within the Bose–Hubbard model, where it is implicitly used that $\epsilon = 0$. NU can therefore be interpreted as the chemical potential of this particular state. The criterion (14) therefore also allows the interpretation that the chemical potential μ should be well below the band gap e_{gap}

$$\frac{\mu}{e_{gap}} \ll 1 \tag{4.17}$$

the criterion (4.16) can also be evaluated using other methods besides the Bose–Hubbard model. We note that the criteria (4.15) and (4.16) place an upper bound on the number of particles, which the authors of Ref. [2] estimated to be in the hundreds of particles for typical parameter values in a three-dimensional setup with just two energy levels below the barrier. In Chap. 6 we will show that even if the criterion (4.17) is satisfied the Bose–Hubbard model can fail to describe the physics by comparison with the exact solution of the many-body Schrödinger equation.

References

1. W. Kohn, Phys. Rev. **115**, 809 (1959)
2. G.J. Milburn, J. Corney, E.M. Wright, D.F. Walls, Phys. Rev. A **55**, 4318 (1997)

3. A. Smerzi, S. Fantoni, S. Giovanazzi, S.R. Shenoy, Phys. Rev. Lett. **79**, 4950 (1997)
4. S. Raghavan, A. Smerzi, S. Fantoni, S.R. Shenoy, Phys. Rev. A **59**, 620 (1999)
5. R. Gati, M.K. Oberthaler, J. Phys. B **40**, R61 (2007)
6. J.C. Eilbeck, P.S. Lomdahl, A.C. Scott, Physica D **16**, 318 (1985)
7. P.A.F. Matthew, B.W. Peter, G. Grinstein, S.F. Daniel, Phys. Rev. B **40**, 546 (1989)
8. D. Jaksch, C. Bruder, J.I. Cirac, C.W. Gardiner, P. Zoller, Phys. Rev. Lett. **81**, 3108 (1998)
9. D. Jaksch, P. Zoller, Ann. Phys. **315**, 52 (2005)
10. L. Pitaevskii, S. Stringari, Bose–Einstein Condensation. (Oxford University Press, Oxford, 2003)
11. C. J. Pethick, H. Smith, Bose–Einstein Condensation in Dilute Gases. 2nd edn. (Cambridge University Press, Cambridge, 2008)

Chapter 5
Reduced Density Matrices and Coherence of Trapped Interacting Bosons

The first- and second-order correlation functions of trapped, interacting Bose–Einstein condensates are investigated numerically on a many-body level from first principles. Correlations in real space and momentum space are treated. The coherence properties are analyzed. The results are obtained by solving the many-body Schrödinger equation. It is shown in an example how many-body effects can be induced by the trap geometry. A generic fragmentation scenario of a condensate is considered. The correlation functions are discussed along a pathway from a single condensate to a fragmented condensate. It is shown that strong correlations can arise from the geometry of the trap, even at weak interaction strengths. The natural orbitals and natural geminals of the system are obtained and discussed. It is shown how the fragmentation of the condensate can be understood in terms of its natural geminals. The many-body results are compared to those of mean-field theory. The best solution within mean-field theory is obtained and the limits in which mean-field theories are valid are determined. In these limits the behavior of the correlation functions is explained within an analytical model. The results of this Chapter were first published in Ref. [1].

5.1 Introduction

The computation of correlation functions in interacting quantum many-body systems is a challenging problem of contemporary physics. Correlations between particles can exist in time, in real space or in momentum space. Since the first experimental realization of Bose–Einstein condensates (BECs) in ultracold atomic gases [2–4], great experimental and theoretical progress has been made in the determination of the coherence and the correlation functions of ultracold bosons. Over the years experiments have measured more and more accurately first, second and to some extent even third-order correlations of trapped BECs, see Refs. [5–12].

Theoretically, the correlation functions of trapped, interacting BECs have been investigated in numerous works, see e.g. Refs. [13–19]. While analytical approaches from first principles are usually restricted to treat homogeneous gases without any

K. Sakmann, *Many-Body Schrödinger Dynamics of Bose–Einstein Condensates*,
Springer Theses, DOI: 10.1007/978-3-642-22866-7_5,

trapping potential, numerical methods can overcome this restriction. It is important to note that the shape of the trapping potential can have a substantial impact on the nature of the quantum state. This is particularly true for issues concerning condensation [20] and fragmentation of Bose systems [21, 22]. For example, the ground state of weakly interacting condensates in harmonic traps is almost fully condensed, while the ground state of double-well potentials can be fragmented or condensed, depending on the height of the barrier, the number of particles and the interaction strength [1, 23–29].

In this Chapter we investigate first- and second-order correlations of trapped, interacting condensates and their coherence properties depending on the trap geometry from first principles. Our results are obtained by solving the many-body Schrödinger equation of the interacting system numerically. From this many-body solution we extract the first- and second-order reduced density matrices which allow us to compute all real and momentum space first- and second-order correlations and in particular the fragmentation of the condensate. For illustration purposes we consider a stationary system in the ground state to show how many-body effects can become dominant when the trap geometry is varied. Here, we are working in a parameter regime where according to the classification scheme given in Refs. [29, 30] and Sect. 2.11 Gross–Pitaevskii theory should be applicable. We stress that our results cannot be described using Gross–Pitaevskii theory.

As a numerical method to solve the interacting many-body problem we use the MCTDHB method [31, 32]. In order to identify true many-body effects in the correlation functions, we compare our many-body results with those based on the most general mean-field approach, which is known as the *best mean-field* (BMF) solution and will be discussed below. A general method to compute this best mean-field solution was developed recently [24–27, 33]. For completeness we compare the many-body results also to the results of Gross–Pitaevskii theory.

In order to understand first- and second-order correlations in an intuitive way, we develop an analytical mean-field model which explains the general structure of our results in those regions where many-body effects can be safely neglected.

5.2 Numerical Methods

The goal of this Chapter is to investigate the first- and second-order correlation functions and the coherence of trapped interacting bosons. In some cases, when the general form of the wave function is known *a priori*, an exact solution can be obtained, either by solving transcendental equations or by exploiting mapping theorems, see e.g. [19, 34–42]. However, in general it is necessary to solve the full many-body Schrödinger equation numerically in an efficient way. Here we employ the MCTDHB method [31, 32] to achieve this goal, see Sect. 3.1 for an explanation. We compare these many-body results with those of Gross–Pitaevskii and best mean-field theory. For an explanation of Gross–Pitaevskii theory and its relation to MCTDHB see Sect. 3.2. For ground states the best mean-field wave function is given by that particular mean-field wave function that minimizes the energy expectation value. The BMF solution *can* coincide with the Gross–Pitaevskii mean-field solution, but this is by no means necessary. In fact, it has been shown [24–27, 33] that the GP mean-field is not always the energetically-lowest mean-field solution. A more detailed account of BMF theory is given in Appendix E and in Refs. [24–27, 33].

5.3 A Generic Example of a Trapped Condensate

5.3.1 General Remarks

In order to examine correlation functions of Bose-condensed systems, we now turn to a specific one-dimensional example. Generally, a trapped condensate will reside in a trapping potential that displays some number of potential maxima and minima. For simplicity we will study the correlation functions of repulsively interacting bosons in a double-well trap. In order to isolate physical effects that are due to the trapping geometry and not to dynamical parameters such as the rate at which the barrier is raised, etc., we restrict our discussion to the ground state at different barrier heights. The restriction to a stationary state allows us to omit the time argument in all physical quantities for the rest of this Chapter. Double-well systems have the interesting property that depending on the height of the barrier and/or the interaction strength, the ground state undergoes a transition from a single to a fragmented condensate [23, 25, 27, 28]. We will show how this transition from a condensed state to a fragmented condensate manifests itself in the correlation functions. The dynamics of a similar system has been investigated recently in the context of a dynamically raised barrier [31].

5.3.2 Trap Parameters and Interaction Strength

Our starting point is the general many-body Hamiltonian given in Eq. 2.6 with a symmetric external potential that consists of a harmonic trap with an additional central barrier of variable height

$$V(x) = \frac{1}{2}x^2 + Ae^{-x^2/2\sigma^2}, \tag{5.1}$$

where A is the height of the potential barrier and $\sigma = 2$ a fixed width. The potential $V(x)$ in Eq. 5.1 approaches $x^2/2$ rapidly as $x \rightarrow \infty$ and is therefore asymptotically homogeneous. As an interaction potential we choose $W(x-x') = \lambda_0\delta(x-x')$ and fix the interparticle interaction at $\lambda_0 = 0.01$. The stationary Schrödinger equation $H\Psi = E\Psi$ is then solved at barrier heights ranging from $A = 0$ to $A = 30$ for $N = 1{,}000$ bosons. In the computations using MCTDHB we restrict the number of orbitals to $M = 2$, yielding a total of $\binom{N+1}{N} = 1{,}001$ variationally optimized permanents. The distinction between variationally optimized and non-optimized permanents is crucial as is demonstrated in an example in Appendix F.

The connection to a real physical Hamiltonian is done as follows. For a given length scale L and a given boson of mass m the unit of energy is $\hbar^2/(mL^2)$ and the unit of the coupling parameter is $\hbar^2/(mL)$. For example if ^{87}Rb is chosen as a boson and $L = 1\,\mu m$ as a length scale then one energy unit equals $(2\pi\hbar)116\,$Hz. With this choice of units the largest barrier height is about $(2\pi\hbar)3.5\,$kHz, with a distance of about $8\,\mu$m between the minima of the trap. Of course other choices can be made.

5.3.3 Condensed State

We begin with a discussion of the ground state energy as a function of the barrier height. The ground state energy per particle of the many-body solution, E_{MCTDHB}/N (blue), is shown in Fig. 5.1(top). E_{MCTDHB}/N increases with the height of the central barrier. The energy differences per particle of the many-body and the BMF solution with respect to the GP solution, $(E_{MCTDHB} - E_{GP})/N$ (blue) and $(E_{BMF} - E_{GP})/N$ (red), are shown in the inset of Fig. 5.1 (top). The energy difference $(E_{MCTDHB} - E_{GP})/N$ is negative, because the interacting system can lower its energy by depleting the condensate. At low barrier heights the GP mean-field is the best mean-field and thus $(E_{BMF} - E_{GP})/N = 0$. A comparison of the energy scales of Fig. 5.1 and its inset reveals that the energy of the many-body solution, the BMF solution and the GP solution are very close at all barrier heights. The nature of the many-body ground state at different barrier heights varies nevertheless very strongly, as shown below.

Figure 5.1 (middle) shows the occupations $n_1^{(1)}/N$ and $n_2^{(1)}/N$ of the first and second natural orbitals of the many-body wave function as a function of the barrier height, computed with MCTDHB. The largest eigenvalue of the first-order RDM, $n_1^{(1)}$, is only restricted by Eq. 2.25 and can therefore take on any value between 0 and N. The dashed lines indicate these upper and lower bounds on $n_1^{(1)}$. At low barrier heights only one natural orbital is significantly occupied. Therefore, we refer to the parameter range $0 \le A \le 13$ as the condensed regime, in accordance with the definition of Penrose and Onsager [20]. The occupation of the second natural orbital is due to the two-body interaction between the particles. However, it remains below 1% for all barrier heights $A \le 13$ and is even below $1‰$ at $A = 0$.

Since $n_1^{(1)} \approx N$ in the condensed regime, the upper and the lower bounds, Eqs. 2.25 and 2.36, on the largest eigenvalue of the second-order RDM, $n_1^{(2)}$, are almost equal. Therefore, $n_1^{(2)}$ is constrained to take on a value very close to $N(N-1)$. Consequently, there can be only one significantly occupied natural geminal. This is confirmed in Fig. 5.1 (bottom), where the natural geminal occupations are shown as a function of the barrier height. For the purpose of describing first- and second-order correlations it is therefore legitimate to approximate the many-body wave function in this regime by a single permanent $|N, 0\rangle$ in which all N bosons occupy the first natural orbital $\alpha_1^{(1)}(x_1)$.

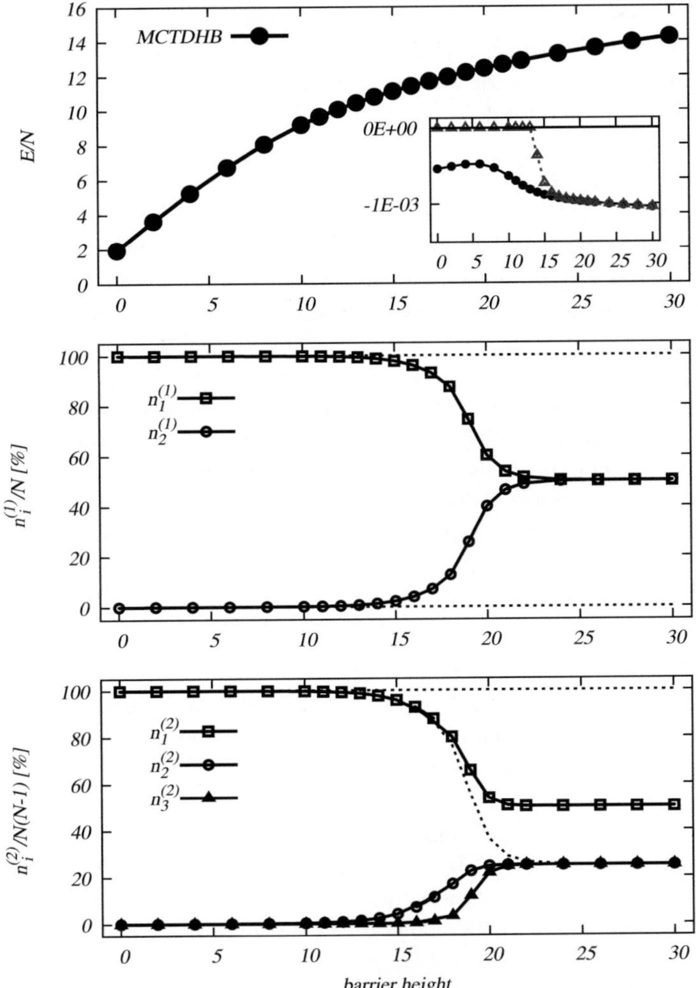

Fig. 5.1 Energy per particle, natural orbital and natural geminal occupations of the ground state of $N = 1{,}000$ bosons at $\lambda_0 = 0.01$ in a harmonic trap with a central barrier. Shown is the dependence on the barrier height. Top: energy per particle E/N of the many-body solution. Inset: energy difference per particle between the best mean-field and the GP solution, $(E_{BMF} - E_{GP})/N$ (*triangles*), and between the many-body and GP solution, $(E_{MCTDHB} - E_{GP})/N$ (*circles*). Middle: the eigenvalues $n_1^{(1)}$ and $n_2^{(1)}$ of the first-order RDM $\rho^{(1)}(x_1|x_1')$. The ground state fragments with increasing barrier height. Bottom: the eigenvalues $n_1^{(2)}$, $n_2^{(2)}$ and $n_3^{(2)}$ of the second-order RDM $\rho^{(2)}(x_1, x_2|x_1', x_2')$. The *dashed lines* in the middle and bottom panel indicate upper and lower bounds on the largest eigenvalue of the first- and second-order RDMs. See text for details. The quantities shown are dimensionless

Note that the natural orbitals and the natural geminals are generally complex functions. However, the ground state wave function is real and hence the natural orbitals and the natural geminals are real functions. The first column of Fig. 5.2

shows the first (red) and the second (blue) natural orbitals of the many-body solution at barrier heights $A = 0$, 13, 19, 24, from top to bottom. The first and the second natural orbitals are symmetric and antisymmetric about the origin, respectively. At $A = 0$ the first natural orbital, $\alpha_1^{(1)}(x_1)$, takes on the shape of a broadened Gaussian, reflecting the repulsive interaction between the particles. The second natural orbital, $\alpha_2^{(1)}(x_1)$, has a higher kinetic energy than the first one due to the node at the center of the trap. Additionally, the second natural orbital forces the particles to occupy regions of the trap where the trapping potential is higher. There is an energy gap between the one-particle energies of the first and second natural orbital. The occupation of the second natural orbital is therefore very small in the purely harmonic trap at the chosen interaction strength.

As the barrier height is varied from $A = 0$ up to $A = 13$, the natural orbitals deform to fit the new shape of the external potential. The central peak of the first natural orbital splits into two maxima which become localized at positions $x_1 = \pm d/2$, where d is the distance between the wells of the external potential.

At the center of the trap, where the barrier is raised, the first natural orbital develops a local minimum in order to minimize the potential energy. The second natural orbital on the other hand has a node at the center of the trap at any barrier height. Its maximum and minimum are localized at the minima of the external potential. As the barrier is raised, the energy gap between the first two natural orbitals decreases. However, the increase of the depletion of the condensate from $n_2^{(1)}/N < 1‰$ at $A = 0$ to $n_2^{(1)}/N \approx 1\%$ at $A = 13$ cannot be explained in this single particle picture. On a single particle level the ground state would be fully condensed at any finite barrier height, i.e. $n_2^{(1)} = 0$. The reason for the observed increase in the depletion lies in the fact that for repulsively interacting many-boson systems in multi-well setups it becomes energetically more favourable to fragment as the barrier between the wells is raised [23, 25, 27, 28, 33]. In some The increase in energy which results from the occupation of orbitals with a higher one-particle energy can be outweighed by a decrease in interaction energy. This effect becomes dominant at barrier heights above $A = 13$, see Sects. 5.3.4 and 5.3.5

The second to fourth column of Fig. 5.2 show (from left to right) the first three natural geminals $\alpha_i^{(2)}(x_1, x_2)$ at the same barrier heights as above. From the numerical many-body simulation we find that the natural geminals in the condensed regime are approximately given by symmetrized products of the natural orbitals:

$$|\alpha_1^{(2)}\rangle = |2, 0\rangle, \quad |\alpha_2^{(2)}\rangle = |1, 1\rangle, \quad |\alpha_3^{(2)}\rangle = |0, 2\rangle, \tag{5.2}$$

where $|m_1, m_2\rangle$ denotes a state with m_1 particles in the first and m_2 particles in the second natural orbital. Only the first natural geminal, $\alpha_1^{(2)}(x_1, x_2)$, is significantly occupied in the condensed regime. Due to the two-body interaction between the particles there is a small occupation of the second and third natural geminal. However, at low barrier heights their occupation is largely suppressed, due to the gap between the single particle energies of the first and second natural orbital. Since the geminals

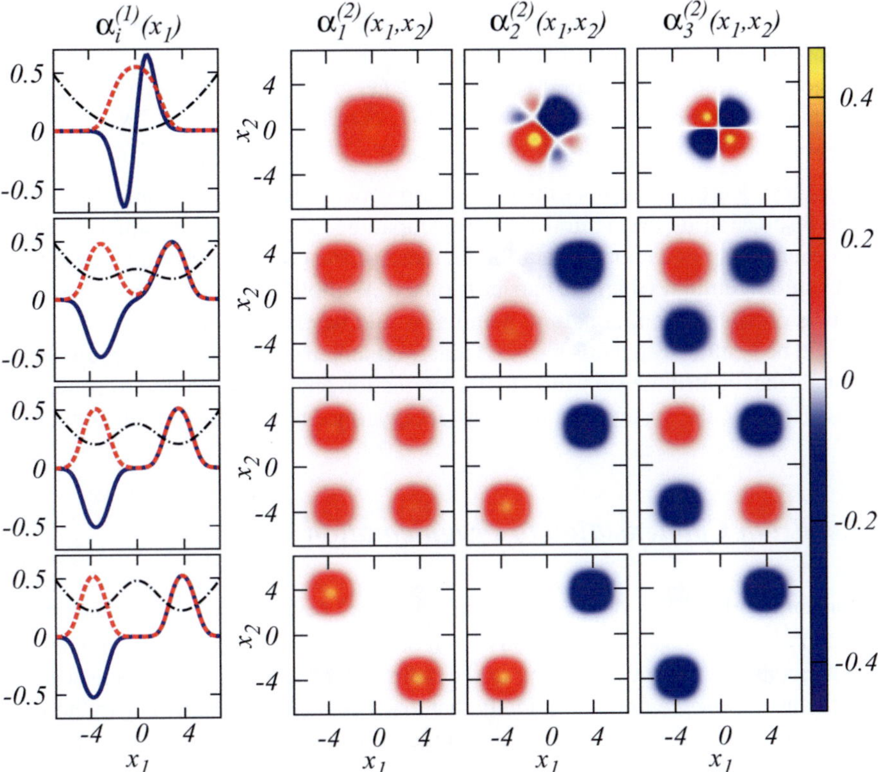

Fig. 5.2 Natural orbitals and geminals at different barrier heights. First column: the natural orbitals $\alpha_1^{(1)}(x_1)$ (*dashed red line*) and $\alpha_2^{(1)}(x_1)$ (*solid blue line*) of the many-body solution at different barrier heights $A = 0, 13, 19, 24$, from top to bottom. The trapping potential is shown as a *dashed-dotted black line* in the first column. The state of the system changes from condensed to fragmented between $A = 13$ and $A = 24$. Second to fourth columns: natural geminals $\alpha_1^{(2)}(x_1, x_2)$, $\alpha_2^{(2)}(x_1, x_2)$ and $\alpha_3^{(2)}(x_1, x_2)$ from left to right at the same barrier heights as above. While the natural orbitals remain qualitatively unchanged during the fragmentation transition, the natural geminals take on their final shapes only when the system becomes fully fragmented. At low barrier heights only one natural geminal is occupied. At high barriers three natural geminals are occupied, see Fig. 5.1. The total energy is minimized by the occupation of a natural geminal that contributes practically nothing to the interaction energy. See text for more details. The quantities shown are dimensionless

$\alpha_2^{(2)}(x_1, x_2)$ and $\alpha_3^{(2)}(x_1, x_2)$ contain the second natural orbital in their expansion, see Eq. 5.2, their occupation increases the total energy at low barrier heights.

Since $n_1^{(2)} \approx N(N-1)$ in the condensed regime, the only substantially contributing natural geminal in the equation for the energy expectation value, Eq. 2.27, is $\alpha_1^{(2)}(x_1, x_2)$. The shape of $\alpha_1^{(2)}(x_1, x_2)$ is therefore particularly interesting. It has four maxima of similar height, located at positions $x_1 = x_2 = \pm d/2$ and $x_1 = -x_2 = \pm d/2$, see the second panel in the second row of Fig. 5.2. Since $\alpha_1^{(2)}(x_1, x_2)$

has peaks on the diagonal at $x_1 = x_2 = \pm d/2$, it contributes to both, the one-particle part and the interaction part of the energy.

In contrast to the first natural geminal, $\alpha_2^{(2)}(x_1, x_2)$ and $\alpha_3^{(2)}(x_1, x_2)$ both exhibit node lines going through the region where the central barrier is raised. As the energy gap between the single particle energies of the natural orbitals $\alpha_1^{(1)}(x_1, x_2)$ and $\alpha_2^{(1)}(x_1, x_2)$ decreases, so does the energy gap between the natural geminals. Similar to the discussion of the natural orbital occupations above, this argument in terms of an energy gap does not explain the increase of the occupation of the second and third natural geminal when the barrier is raised. Without interactions the occupation numbers of all but the first natural geminal would be exactly zero.

It will be demonstrated in Sect. 5.3.5 that fragmented states allow the occupation of geminals that contribute very little to the interaction energy as opposed to condensed states. Thereby, the system can lower its energy, once the barrier is high enough.

5.3.4 From Condensation to Fragmentation

At barrier heights $13 < A < 24$, one finds that the occupation of the second natural orbital $n_2^{(1)}/N$ increases continuously from below 1 to almost 50%. The condensate fragments in this regime according to the definition of fragmented condensates [21, 22]. In this regime we find numerically that many permanents contribute to the wave function and, therefore, we refer to the range of barrier heights $13 < A < 24$ as the many-body regime.

Along with the natural orbital occupations the natural geminal occupations change as well. Three natural geminals become occupied with increasing barrier height, see Fig. 5.1. In the many-body regime, the upper and lower bounds, Eqs. 2.25 and 2.36, on the largest eigenvalue of the second-order RDM, $n_1^{(2)}$, no longer restrict $n_1^{(2)}$ to a narrow region. In fact, $n_1^{(2)}$ takes on a value somewhere in between these bounds.

This onset of fragmentation manifests itself also in the BMF solution which jumps from a GP type permanent $|N, 0\rangle$ in the condensed regime to a fully fragmented solution of the form $|N/2, N/2\rangle$. Note that already at barrier heights $A \geq 14$ this fragmented solution is lower in energy than a GP type permanent. At this barrier height the many-body solution is only slightly depleted, see Fig. 5.2.

If we compare the natural orbitals in Fig. 5.2 at barrier height $A = 19$ with those at $A = 13$, we note that they look very similar, apart from the fact that the peaks are slightly farther apart, and the first natural orbital is closer to zero at the center of the trap. The energies of the first two natural orbitals are almost degenerate and the total energy is minimized by the occupation of both natural orbitals. Without interactions the system would remain in a condensed state, since the single particle energies of the first and the second natural orbitals remain separated at any finite barrier height. Note that in the absence of interactions the natural orbitals are the eigenfunctions of $\hat{h}$. However, as we noted in Sect. 5.3, a system of repulsively interacting bosons in multi-well traps can lower its energy by occupying several natural orbitals, once the barrier is high enough [25–28, 33]. This is precisely the reason for the observed onset of fragmentation.

In the many-body regime the natural geminals are no longer symmetrized products of the natural orbitals. If we compare $\alpha_1^{(2)}(x_1, x_2)$ in Fig. 5.2 at barrier heights $A = 13$ and $A = 19$, we see that the peaks on the diagonal at $x_1 = x_2 = \pm d/2$ decrease, whilst the peaks on the off-diagonal at $x_1 = -x_2 = \pm d/2$ increase when the barrier is raised. The opposite is true for the third natural geminal $\alpha_3^{(2)}(x_1, x_2)$: the off-diagonal maxima at $x_1 = -x_2 = \pm d/2$ have decreased, whilst the diagonal minima at $x_1 = x_2 = d/2$ are now more negative. On the other hand, the second natural geminal, $\alpha_2^{(2)}(x_1, x_2)$, is still well approximated by a symmetrized product of the first and second natural orbital. The behavior of the natural geminals is qualitatively different from that displayed by the natural orbitals. In contrast to the natural orbitals, the natural geminals *do* change their shape during the fragmentation transition. They only obtain their final forms, when the fragmentation transition is completed, see Fig. 5.2 and Sect. 5.3.5.

5.3.5 Fully Fragmented State

When the central barrier is raised to values $A \geq 24$, the two parts of the condensate become truly independent. The natural orbital occupations approach $n_1^{(1)} = n_2^{(1)} = N/2$, which reflects the fact that the energies associated with the first and second natural orbitals degenerate at infinite barrier heights. The many body wave function can then be adequately approximated by a single permanent of the form $|N/2, N/2\rangle$, i.e. with equal numbers of particles in the first and the second natural orbitals. Therefore, we refer to barrier heights $A \geq 24$ as the fully fragmented regime. The additional energy, necessary for the occupation of the second natural orbital, is outweighed by a lower interaction energy. Note that this final form of the wave function is anticipated by the BMF solution at barrier heights $A \geq 14$.

The natural geminal occupations approach

$$n_1^{(2)} = N(N/2), \quad n_2^{(2)} = n_3^{(2)} = N/2(N/2 - 1) \tag{5.3}$$

in the fully fragmented regime. These are the values that follow from the BMF solution.

It is only at barrier heights $A \geq 24$ that the natural geminals take on their final shapes, compare the third and fourth rows of Fig. 5.2. If we expand the natural geminals in the basis of natural orbitals at these barrier heights, we find that

$$\begin{aligned}
|\alpha_1^{(2)}\rangle &= \frac{1}{\sqrt{2}}\Big(|2, 0\rangle - |0, 2\rangle\Big) \\
|\alpha_2^{(2)}\rangle &= |1, 1\rangle \\
|\alpha_3^{(2)}\rangle &= \frac{1}{\sqrt{2}}\Big(|2, 0\rangle + |0, 2\rangle\Big)
\end{aligned} \tag{5.4}$$

holds to a very good approximation. The first and third natural geminals have equal contributions coming from the first and the second natural orbitals. The question arises, why their occupations are different, about 50 and 25%, respectively. Subtracting the permanents $|2,0\rangle$ and $|0,2\rangle$ from one another yields a geminal which is localized on the *off-diagonal*, see $\alpha_1^{(2)}(x_1,x_2)$ in Fig. 5.2 at $A=24$. Adding the permanents $|2,0\rangle$ and $|0,2\rangle$ yields a geminal which is localized on the *diagonal*, see $\alpha_3^{(2)}(x_1,x_2)$ in Fig. 5.2 at $A=24$. It is easy to see from the shape of the natural geminals in the fourth row of Fig. 5.2 that the integrals over the one-body part in Eq. 2.27 are approximately the same for each of the natural geminals. Given the occupations in Eq. 5.3, the first natural geminal contributes about one half of the one-body energy, whereas the second and the third natural geminal contribute about a fourth each. The situation is different for the two-body part of the Hamiltonian. Since $\alpha_2^{(2)}(x_1,x_2)$ and $\alpha_3^{(2)}(x_1,x_2)$ are localized on the diagonal, they *do* contribute to the interaction energy. In contrast, $\alpha_1^{(2)}(x_1,x_2)$ is almost zero at coordinate values $x_1\approx x_2$ and, due to the contact interaction $W(x-x')=\lambda_0\delta(x-x')$, it practically does not contribute to the interaction energy. At high barriers a fragmented state allows the system to lower its energy through the occupation of a natural geminal which is localized on the off-diagonal.

By evaluating the Lieb–Liniger parameter γ as defined in Eq. 2.43 we find that $\gamma\approx 5\times 10^{-5}$ at all barrier heights and according to the classification scheme in Sect. 2.11 the system should have been deep in the 1D Thomas-Fermi limit and hence no fragmentation should have occured. This was obviously not the case, the system fragmented and we conclude that even at a particle number of $N=1{,}000$ finite size effects can still dictate the nature of the quantum state.

We would like to make a remark on the validity of the present MCTDHB computation for high barriers. For high barriers the whole system can be considered as composed of two separate condensates. To describe the depletion of each condensate it would be necessary to employ $M=4$ orbitals. We use only $M=2$ orbitals in the many-body computation and cannot describe this depletion. We justify the use of $M=2$ orbitals by noting that at $A=0$ the system is almost fully condensed, and the depletion can be safely neglected, see Fig. 5.1. Therefore, we assume that the depletion of each of the two condensates can be neglected, when the barrier is very high. This claim is supported by a computation that we carried out in the harmonic trap at the same interaction strength $\lambda_0=0.01$ for 500 particles. The depletion was found to be even less than for $N=1{,}000$ particles.

5.4 First-Order Correlations

5.4.1 General Analytical Considerations

We now describe the first-order correlations in an analytical mean-field model for the two limiting cases of a condensed and a fully fragmented system. In these cases mean-field theory has been shown to be well applicable, see Sect. 5.3. For our purposes the exact shape of the natural orbitals $\alpha_1^{(1)}(x_1)$ and $\alpha_2^{(1)}(x_1)$ is unimportant. Consider a normalized one-particle function, $\Phi(x)$, which is localized at the origin. $\Phi(x)$ may vary in shape, but is always assumed to resemble a Gaussian. Similarly, we define translated copies $\Phi_1(x) = \Phi(x + d/2)$ and $\Phi_2(x) = \Phi(x - d/2)$ of $\Phi(x)$, where the previously defined distance d between the minima of the potential wells is taken to be large enough to set products of the form $\Phi_1(x)\Phi_2(x)$ to zero. Since Φ is localized in some region around the origin, Φ_1 is localized in a region L to the left and Φ_2 in a region R to the right of the origin.

5.4.1.1 Condensed State

In the condensed regime, $0 \leq A \leq 13$, only one natural orbital, $\alpha_1^{(1)}(x_1)$, is significantly occupied. Therefore, we approximate the first-order reduced density operator of the system by that of a condensed state $|N, 0\rangle$

$$\hat{\rho}^{(1)}_{|N,0\rangle} = N|\alpha_1^{(1)}\rangle\langle\alpha_1^{(1)}|. \tag{5.5}$$

It then follows from Eq. 2.29 that

$$|g^{(1)}_{|N,0\rangle}(x_1', x_1)|^2 = 1. \tag{5.6}$$

At zero barrier height, the first natural orbital is a Gaussian, broadened by interactions. Therefore, we write $\alpha_1^{(1)}(x_1) = \Phi(x_1)$, and hence the one-particle density distribution and the one-particle momentum distribution are of the form

$$\rho^{(1)}_{|N,0\rangle}(x_1|x_1) = N|\Phi(x_1)|^2, \tag{5.7}$$

$$\rho^{(1)}_{|N,0\rangle}(k_1|k_1) = N|\Phi(k_1)|^2. \tag{5.8}$$

Since $\Phi(x_1)$ is a broadened Gaussian, its Fourier transform $\Phi(k_1)$ is also close to a Gaussian, but narrower in comparison to a non-interacting system. The momentum distribution of the repulsively interacting system in the harmonic trap is therefore narrower than that of a non-interacting system.

We now turn to the case corresponding to $A \approx 13$, where the system is still condensed, but the first two natural orbitals are spread out over the two wells. We model the natural orbitals by

$$\alpha_1^{(1)}(x_1) = \frac{1}{\sqrt{2}}\Big[\Phi_1(x_1) + \Phi_2(x_1)\Big], \quad \alpha_2^{(1)}(x_1) = \frac{1}{\sqrt{2}}\Big[\Phi_1(x_1) - \Phi_2(x_1)\Big]. \tag{5.9}$$

In this case one obtains [43]:

$$\rho^{(1)}_{|N,0\rangle}(x_1|x_1) = \frac{N}{2}|\Phi_1(x_1)|^2 + \frac{N}{2}|\Phi_2(x_1)|^2, \tag{5.10}$$

$$\rho^{(1)}_{|N,0\rangle}(k_1|k_1) = N[1 + \cos(k_1 d)]|\Phi(k_1)|^2 \tag{5.11}$$

for the density and the momentum distribution. We note that the one-particle momentum distribution displays an oscillatory pattern in momentum space at a period which is determined by the separation d of the centers of the two wells.

5.4.1.2 Fully Fragmented State

In the true many-body regime, $13 < A < 24$, where many permanents contribute to the wave function, a mean-field model is bound to fail. However, in the fully fragmented regime it is possible to consider the whole system as two separate condensates, and hence a mean-field description is again applicable. Therefore, we now turn to the case corresponding to $A \geq 24$, where the system is fully fragmented and the many-body state is given by $|N/2, N/2\rangle$. The first-order density operator then reads:

$$\hat{\rho}^{(1)}_{|N/2,N/2\rangle} = \frac{N}{2}|\alpha^{(1)}_1\rangle\langle\alpha^{(1)}_1| + \frac{N}{2}|\alpha^{(1)}_2\rangle\langle\alpha^{(1)}_2|. \tag{5.12}$$

Since the natural orbitals remain qualitatively unchanged during the fragmentation transition, we approximate $\alpha^{(1)}_1(x_1)$ and $\alpha^{(1)}_2(x_1)$ by Eqs. 5.9 and obtain for the density and the momentum distribution [43]:

$$\rho^{(1)}_{|N/2,N/2\rangle}(x_1|x_1) = \frac{N}{2}|\Phi_1(x_1)|^2 + \frac{N}{2}|\Phi_2(x_1)|^2, \tag{5.13}$$

$$\rho^{(1)}_{|N/2,N/2\rangle}(k_1|k_1) = N|\Phi(k_1)|^2. \tag{5.14}$$

We note that the one-particle momentum distribution of independent condensates does not contain an oscillatory component and is identical to the momentum distribution of a single, localized condensate of N particles within this model, see Eq. 5.8. For the normalized first-order correlation function one finds

$$|g^{(1)}_{|N/2,N/2\rangle}(x'_1, x_1)|^2 = \begin{cases} 1 & \text{if } x_1,\ \ x'_1 \in L \text{or} x_1,\ \ x'_1 \in R, \\ 0 & \text{otherwise.} \end{cases} \tag{5.15}$$

Whereas the state $|N, 0\rangle$ is fully first-order coherent, the fragmented state $|N/2, N/2\rangle$ is only first-order coherent in a restricted and generally disconnected region. Each of the two condensates is first-order coherent, but the mutual coherence which is present in the condensed regime is lost.

5.4.2 Numerical Results

We now turn to the discussion of first-order correlations. In particular, we are interested in effects that are due to the true many-body nature of the wave function. Along with our many-body results we plot the corresponding results of the BMF solution. From the discussion in Sect. 5.3 it is clear that we expect many-body effects to occur during the fragmentation transition at barrier heights $13 < A < 24$. In the condensed and in the fully fragmented regime we expect that the many-body results are well approximated by those of the BMF solution. In these cases we can understand the structure of the correlation functions on the basis of the analytical mean-field model of Sect. 5.4.1.

The first column of Fig. 5.3 shows the one-particle density distribution $\rho^{(1)}(x_1|x_1)$ of the many-body solution (blue line) and that of the BMF solution (red line with triangles) at the barrier heights $A = 0$, 13, 19, 24, from top to bottom. It is remarkable that the one-particle densities obtained from either the many-body wave function or the BMF solution give results that cannot be distinguished from one another at *any* barrier height.

In a purely harmonic trap, $A = 0$, the one-particle density takes on the form of an interaction-broadened Gaussian. At higher barriers, the density splits into two parts that are localized in each of the wells. At $A = 13$, the one-particle density has developed two separated peaks. Note that the system is still in the condensed regime at this barrier height and must be considered a single condensate, despite the spatial separation between the two peaks.

When the central barrier is raised further to values $13 < A < 24$ the condensate fragments, see Fig. 5.1. At a barrier height of $A = 19$ the system is halfway on its way from a condensed to a fully fragmented condensate. Many permanents contribute to the many-body wave function and one may wonder how this transition manifests itself in observable quantities. However, apart from a small shift of the center of the two peaks and a reduction at the center of the trap, $\rho^{(1)}(x_1|x_1)$ remains largely unaffected by this transition. If the barrier is raised further to $A = 24$, the fragmentation transition is largely completed. Also during the transition from a true many-body state to a fully fragmented state there is no visible indication of this transition in the one-particle density.

The second column of Fig. 5.3 shows the one-particle momentum distribution $\rho^{(1)}(k_1|k_1)$ at the same barrier heights as before. At $A = 0$, the one-particle momentum distribution is given by a squeezed Gaussian, in agreement with Eq. 5.8. At $A = 13$ the one-particle momentum distribution has developed an oscillatory pattern, typical of a single condensate spread out over two wells. The structure of $\rho^{(1)}(k_1|k_1)$ is well reproduced by Eq. 5.11 of the analytical mean-field model. Up to this barrier height the BMF solution is almost identical to the many-body wave function, and therefore the respective momentum distributions are indistinguishable, see the two upper panels in the second column of Fig. 5.3.

When the system enters the many-body regime, $13 < A < 24$, the momentum distribution of the many-body solution deforms to a Gaussian-like envelope, modulated by an oscillatory part. The BMF momentum distribution, on the other hand, already takes on the form characteristic of two separate condensates. It agrees with the prediction of Eq. 5.14, which is clearly different from the many-body result. This merely reflects the fact that the many-body wave function is inaccessible to mean-field methods in the many-body regime.

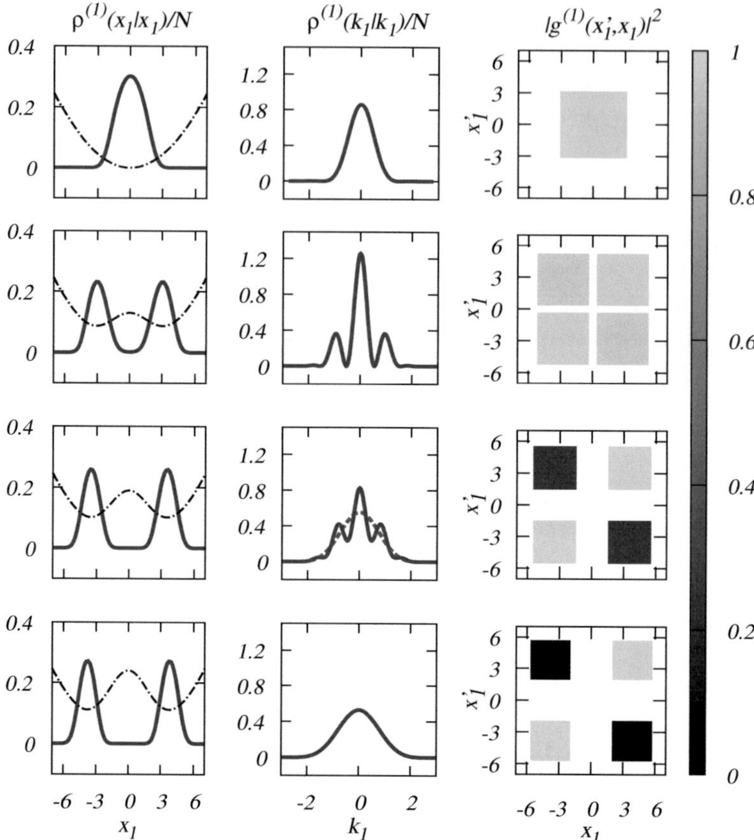

Fig. 5.3 Density distribution, momentum distribution and first-order coherence. The first two columns show the one-particle density $\rho^{(1)}(x_1|x_1)/N$ and the one-particle momentum distribution $\rho^{(1)}(k_1|k_1)/N$ of the many-body solution (*solid lines*) and of the BMF solution (*dashed line*), respectively. The trapping potential is shown as a *dashed-dotted* line. From top to bottom the height of the central barrier is $A = 0$, 13, 19, 24. The BMF result agrees well with the many-body result for a large range of barrier heights. Only at $A = 19$, in the many-body regime, deviations are visible in the momentum distribution. See text for details. The third column shows the absolute value squared of the normalized first-order correlation function $|g^{(1)}(x_1', x_1)|^2$ at the same barrier heights. An initially coherent condensate splits into two separate condensates which are no longer mutually coherent. Only the coherence within each of the wells is preserved. The quantities shown are dimensionless

When the state becomes fully fragmented at $A = 24$, the many-body momentum distribution and the BMF momentum distribution become indistinguishable again, consistent with an explanation in terms of two independent condensates, see Eq. 5.14. Compared to $\rho^{(1)}(k_1|k_1)$ at $A = 0$, the momentum distribution is broader at $A = 24$, because the density distribution in each of the two wells is narrower than that in the harmonic trap.

The third column of Fig. 5.3 shows the absolute value squared of the normalized first-order correlation function $|g^{(1)}(x_1', x_1)|^2$ of the many-body solution only. Here and in all following graphs of correlation functions we restrict the plotted region by a simple rule. To avoid analyzing correlations in regions of space where the density is essentially zero, we plot the respective correlation function only in regions where the density is larger than 1% of the maximum value of the density in the entire space. We apply the same rule also in momentum space.

At zero barrier height $|g^{(1)}(x_1', x_1)|^2$ is very close to one in the region where the density is localized. The system is first-order coherent to a very good approximation and the mean-field formula Eq. 5.6 applies. As the barrier is raised to $A = 13$ the coherence between the two peaks, e.g. at $x_1 = -x_1'$, is slightly decreased, while the coherence within each of the peaks is preserved. Note that the density at the center of the trap is already below 1% of the maximal value in this case. Despite this separation the system remains largely condensed, but deviations from Eq. 5.6 are visible. If the barrier is raised further to $A = 19$, the coherence of the system on the off-diagonal decreases quickly. Although the bosons in each well remain coherent among each other, the overall system is only partially coherent. At barrier heights $A \geq 24$, the coherence between the two wells is entirely lost. This is also the scenario that the BMF solution anticipates, see Eq. 5.15.

It is remarkable that not only the density, but also the momentum distribution obtained within mean-field theory agree so well with the many-body result, when the system is not in the true many-body regime. This would not be the case if we had restricted the mean-field approach to the GP equation, as we will show now.

Up to barrier heights $A = 13$ the many-body system is condensed, and the BMF solution coincides with the GP solution. The BMF, and therefore also the GP solution provide good approximations to the interacting many-body system. Above $A = 13$ the results obtained with the GP mean-field become qualitatively wrong as the barrier is raised. To illustrate this point, we plot the GP results corresponding to those of Fig. 5.3 at barrier heights $A = 19$ (top) and $A = 24$ (bottom) in Fig. 5.4. A comparison of the respective one-particle densities, shown in the first column of Figs. 5.3 and 5.4, reveals no visible difference. The GP mean-field reproduces the density distribution at all barrier heights correctly. However, the GP solution fails at the description of the momentum distribution and the normalized first-order correlation function, compare the second and third columns of Figs. 5.3 and 5.4 at the same barrier heights. The reason for the failure of the GP mean-field is the assumption that all bosons occupy the same orbital. It is by construction incapable to describe fragmented condensates.

5.5 Second-Order Correlations

5.5.1 General Analytical Considerations

In this subsection we extend the analytical mean-field model of Sect. 5.4.1 to describe second-order correlations.

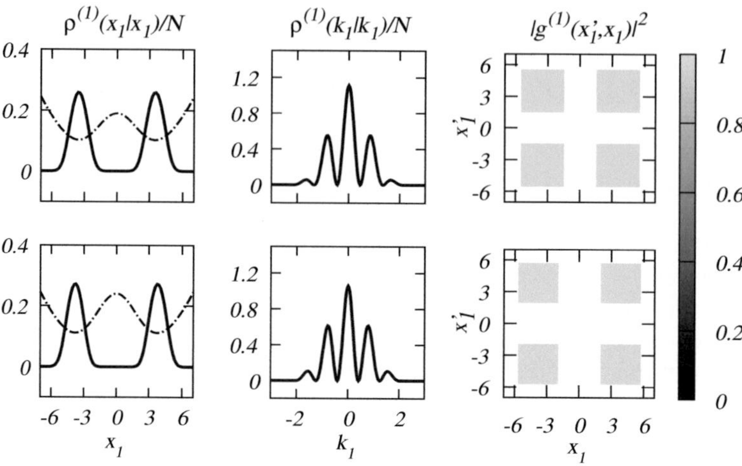

Fig. 5.4 Density distribution, momentum distribution and first-order coherence obtained by using the GP equation for high barriers. The first two columns show the GP one-particle density $\rho^{(1)}(x_1|x_1)/N$ (*left*) and the GP one-particle momentum distribution $\rho^{(1)}(k_1|k_1)/N$ (*middle*) at barrier heights $A = 19$ and $A = 24$ (*solid lines*). In the first column the trapping potential is also shown (*dashed-dotted line*). The GP equation models the density well, but fails at the computation of the momentum distribution, compare with Fig. 5.3. The third column shows the absolute value squared of the normalized first-order correlation function $|g^{(1)}(x_1', x_1)|^2$ computed with the GP equation at the same barrier heights. The normalized first-order correlation function is incorrectly described by the solution of the GP equation. The quantities shown are dimensionless

5.5.1.1 Condensed State

We found in Sect. 5.3.3 that only one natural geminal is significantly occupied in the condensed regime, where the many-body state is approximately given by a single permanent in which all bosons occupy the same single particle state. Therefore, we approximate the second-order reduced density operator in the condensed regime by that of the state $|N, 0\rangle$:

$$\hat{\rho}^{(2)}_{|N,0\rangle} = N(N-1)|\alpha_1^{(2)}\rangle\langle\alpha_1^{(2)}|, \tag{5.16}$$

where $\alpha_1^{(2)}(x_1, x_2) = \alpha_1^{(1)}(x_1)\alpha_1^{(1)}(x_2)$ is the permanent in which two bosons reside in the first natural orbital $\alpha_1^{(1)}$. For the condensed state $|N, 0\rangle$ one finds that up to corrections of order $\mathcal{O}(1/N)$ the state is second-order coherent:

$$g^{(2)}_{|N,0\rangle}(x_1, x_2, x_1, x_2) = 1 - \frac{1}{N}, \tag{5.17}$$

$$g^{(2)}_{|N,0\rangle}(k_1, k_2, k_1, k_2) = 1 - \frac{1}{N}. \tag{5.18}$$

Thus, there are practically no two-body correlations if $N \gg 1$. At zero barrier height the first natural orbital takes on the shape of a broadened Gaussian, $\alpha_1^{(1)}(x_1) =$

$\Phi(x_1)$, where $\Phi(x)$ is defined in Sect. 5.1 The first natural geminal then reads $\alpha_1^{(2)}(x_1, x_2) = \Phi(x_1)\Phi(x_2)$. It follows that the two-particle density and the two-particle momentum distribution factorize up to corrections of order $\mathcal{O}(1/N)$ into products of the respective one-particle distributions:

$$\rho^{(2)}_{|N,0\rangle}(x_1, x_2|x_1, x_2) = N(N-1)|\Phi(x_1)|^2|\Phi(x_2)|^2, \tag{5.19}$$

$$\rho^{(2)}_{|N,0\rangle}(k_1, k_2|k_1, k_2) = N(N-1)|\Phi(k_1)|^2|\Phi(k_2)|^2. \tag{5.20}$$

At the barrier height $A = 13$, the system is condensed but spread out over the two wells. Then, using Eq. 5.9 to approximate $\alpha_1^{(1)}$, we find

$$\begin{aligned}\rho^{(2)}_{|N,0\rangle}(x_1, x_2|x_1, x_2) = \frac{N(N-1)}{4}\Big[&|\Phi_1(x_1)\Phi_1(x_2)|^2 + |\Phi_1(x_1)\Phi_2(x_2)|^2 \\ &+ |\Phi_2(x_1)\Phi_1(x_2)|^2 + |\Phi_2(x_1)\Phi_2(x_2)|^2\Big],\end{aligned} \tag{5.21}$$

$$\begin{aligned}\rho^{(2)}_{|N,0\rangle}(k_1, k_2|k_1, k_2) = N(N-1)&[1+\cos(k_1 d)][1+\cos(k_2 d)] \\ &\times |\Phi(k_1)\Phi(k_2)|^2\end{aligned} \tag{5.22}$$

for the two-particle density and the two-particle momentum distribution. Apart from a correction of order $\mathcal{O}(1/N)$, the two-particle density and the two-particle momentum distribution are again products of the respective one-particle distributions.

5.5.1.2 Fully Fragmented State

In Sect. 5.3.5 we found that three natural geminals are occupied in the fully fragmented regime, see Eq. 5.3. The occupations of Eq. 5.3 hold exactly for a state of the form $|N/2, N/2\rangle$. Therefore, we approximate the second-order reduced density operator in the fully fragmented regime by that of the state $|N/2, N/2\rangle$:

$$\begin{aligned}\hat{\rho}^{(2)}_{|N/2,N/2\rangle} = N\frac{N}{2}|\alpha_1^{(2)}\rangle\langle\alpha_1^{(2)}| &+ \frac{N}{2}\left(\frac{N}{2}-1\right)|\alpha_2^{(2)}\rangle\langle\alpha_2^{(2)}| \\ &+ \frac{N}{2}\left(\frac{N}{2}-1\right)|\alpha_3^{(2)}\rangle\langle\alpha_3^{(2)}|,\end{aligned} \tag{5.23}$$

where the natural geminals $|\alpha_i^{(2)}\rangle$ are given by Eq. 5.4. In contrast to the condensed state, the normalized second-order correlation function of the fully fragmented state has a more complicated structure due to the different terms contributing to Eq. 5.23. We approximate the natural geminals using Eq. 5.9 and find

$$\rho^{(2)}_{|N/2,N/2\rangle}(x_1,x_2|x_1,x_2) = \frac{N}{2}\left(\frac{N}{2}-1\right)\left[|\Phi_1(x_1)\Phi_1(x_2)|^2 + |\Phi_2(x_1)\Phi_2(x_2)|^2\right] + \frac{N}{2}\frac{N}{2}\left[|\Phi_1(x_1)\Phi_2(x_2)|^2 + |\Phi_2(x_1)\Phi_1(x_2)|^2\right] \tag{5.24}$$

and

$$\rho^{(2)}_{|N/2,N/2\rangle}(k_1,k_2|k_1,k_2) = N(N-1)\left(1+\frac{N}{N-1}\frac{\cos[(k_1-k_2)d]}{2}\right)|\Phi(k_1)\Phi(k_2)|^2 \tag{5.25}$$

for the two-particle density and the two-particle momentum distribution. This representation allows us to identify the first two terms in Eq. 5.23 as contributions coming from two separate condensates of $N/2$ bosons each, with condensate wave functions $\Phi_1(x_1)$ and $\Phi_2(x_1)$. The third term in Eq. 5.23 is due to the fact that the bosons in the two separated condensates are identical particles. For the normalized second-order correlation function one finds:

$$g^{(2)}_{|N/2,N/2\rangle}(x_1,x_2,x_1,x_2) = \begin{cases} 1-\frac{2}{N} & \text{if } x_1, \quad x_2 \in L \text{ or } x_1, \quad x_2 \in R \\ 1 & \text{otherwise,} \end{cases} \tag{5.26}$$

which mimics a high degree of second-order coherence. However, when $g^{(2)}$ is evaluated on the diagonal in momentum space, one finds

$$g^{(2)}_{|N/2,N/2\rangle}(k_1,k_2,k_1,k_2) = \left(1-\frac{1}{N}\right)\left(1+\frac{N}{N-1}\frac{\cos[(k_1-k_2)d]}{2}\right), \tag{5.27}$$

which displays an oscillatory behavior and deviates significantly from a uniform value of one. Hence the system is clearly not coherent, see Sect. 2.5. The fact that $g^{(2)}_{|N/2,N/2\rangle}(k_1,k_2,k_1,k_2)$ oscillates while $g^{(2)}_{|N/2,N/2\rangle}(x_1,x_2,x_1,x_2)$ is almost constant, illustrates the necessity to examine both quantities in order to quantify second-order spatial coherence. A description of second-order correlations in terms of $\rho^{(2)}(x_1,x_2|x_1,x_2)$ and $g^{(2)}(x_1,x_2,x_1,x_2)$ alone is incomplete, and $\rho^{(2)}(k_1,k_2|k_1,k_2)$ and $g^{(2)}(k_1,k_2,k_1,k_2)$ have to be taken into account. Although this may seem obvious in the present case of a fully fragmented state, this reduction of coherence is more intricate in a state which is only partially fragmented, see following subsection. Whether $g^{(2)}(x_1,x_2,x_1,x_2)$ and $g^{(2)}(k_1,k_2,k_1,k_2)$ together suffice to characterize second-order coherence (possibly up to a phase factor) is a matter of further study.

5.5.2 Numerical Results

In this subsection we discuss the second-order correlations of the many-body solution. We compare the results to those of the BMF solution. When mean-field theory gives a good approximation to the many-body results, we also compare with the analytical mean-field model of Sect. 5.5.1

The first two columns of Fig. 5.5 show the two-particle density $\rho^{(2)}(x_1, x_2|x_1, x_2)$ of the many-body (left) and BMF (right) solutions at barrier heights $A = 0$, 13, 19, 24, from top to bottom. At zero barrier height $\rho^{(2)}(x_1, x_2|x_1, x_2)$ is localized at the center of the trap. The two-particle density factorizes approximately into a product of the one-particle densities: $\rho^{(2)}(x_1, x_2|x_1, x_2) \approx \rho^{(1)}(x_1|x_1)\rho^{(1)}(x_2|x_2)$. This remains true up to barrier heights $A = 13$, where the condensate is spread out over the two wells. The BMF result approximates the many-body result well in the condensed regime, and the structure of $\rho^{(2)}(x_1, x_2|x_1, x_2)$ is that of Eqs. 5.19 and 5.21 at barrier heights $A = 0$ and $A = 13$, respectively.

When the barrier is raised further to $A = 19$, the system fragments. Many permanents contribute to the wave function in this regime and there is no simple formula that relates the occupations of the natural orbitals to the two-particle density. Similar to the one-particle density, described in Sect. 5.4.2, the two-particle density seems to take no notice of the transition from a single to a fragmented condensate. It remains practically unchanged during the transition, apart from a slight shift of the peaks away from each other as the barrier is raised

At even higher barriers, $A \geq 24$, the many-body state becomes fully fragmented and the wave function approaches $|N/2, N/2\rangle$. In this limit it is again possible to describe the two-particle density on a mean-field level. Therefore, the analytical results of Sect. 5.5.1 for the fully fragmented state should apply. In fact, the structure of $\rho^{(2)}(x_1, x_2|x_1, x_2)$ in the fully fragmented regime is that predicted by Eq. 5.24.

The two-particle density of the condensed state just below the fragmentation transition and of the fully fragmented state above the fragmentation transition cannot be distinguished. It is easily verified, that Eqs. 5.21 and 5.24 give rise to the same two-particle density profile up to corrections of order $\mathcal{O}(1/N)$.

In contrast, the fragmentation transition is clearly visible in the two-particle momentum distribution. In the third and fourth columns of Fig. 5.5 the two-particle momentum distribution $\rho^{(2)}(k_1, k_2|k_1, k_2)$ of the many-body (left) and BMF (right) wave function are shown.

In the condensed regime the two-particle momentum distribution is approximately given by the product of one-particle momentum distributions of a single condensate. This agrees with the analytical predictions of Eq. 5.20 at barrier height $A = 0$ and Eq. 5.22 at $A = 13$. The mean-field picture is appropriate here.

In the many-body regime the two-particle momentum distribution $\rho^{(2)}(k_1, k_2|k_1, k_2)$ contains contributions from many permanents. The resulting $\rho^{(2)}(k_1, k_2|k_1, k_2)$ has a structure that lies somewhat in between the two results, Eqs. 5.22 and 5.24, obtained within the analytical mean-field model. The BMF solution is fully fragmented and does not provide an accurate approximation to the many-body two-particle momentum distribution in this regime, see Fig. 5.5, third and fourth columns in the third row from above.

When the barrier is raised to $A = 24$, the many-body state becomes fully fragmented and the mean-field picture is again applicable. The pattern of a single coherent

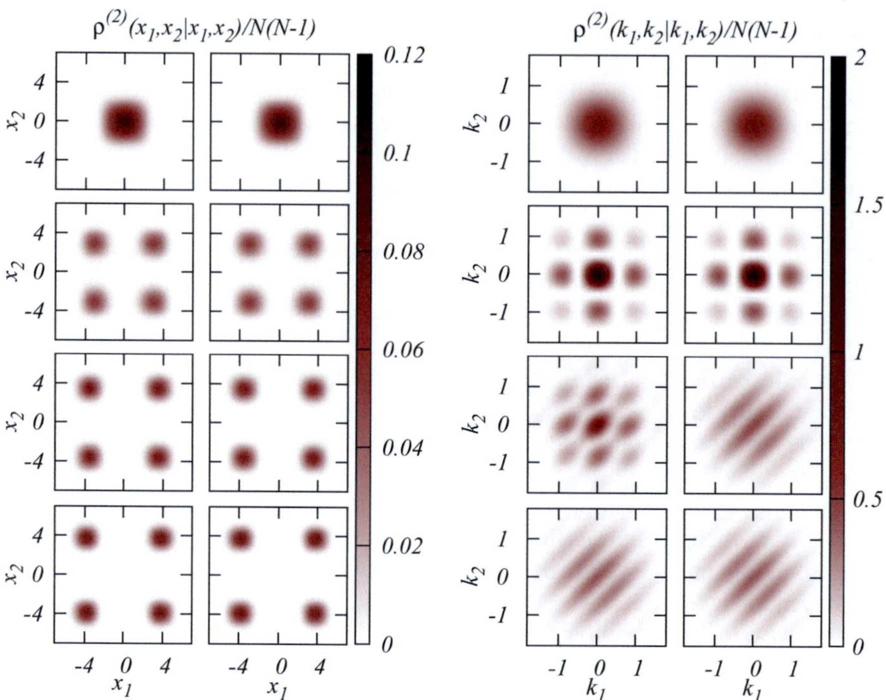

Fig. 5.5 Two-particle density and two-particle momentum distribution at different barrier heights. The first two columns (*from left to right*) show the two-particle density $\rho^{(2)}(x_1, x_2|x_1, x_2)/N(N-1)$ of the many-body (*left*) and BMF (*right*) wave function for the barrier heights $A = 0, 13, 19, 24$, from top to bottom. At low barrier heights ($A = 0, 13$) the system is condensed, and the two-particle density factorizes into a product of the one-particle densities. At higher barriers ($A = 19, 24$), the system fragments and the two-particle density does not factorize into a product of the one-particle densities. The fragmentation transition is not visible in the two-particle density. The results of the many-body and BMF wave function cannot be distinguished at any barrier height. The third and fourth column show the two-particle momentum distribution $\rho^{(2)}(k_1, k_2|k_1, k_2)/N(N-1)$ of the many-body (*left*) and BMF (*right*) solution at the same barrier heights as above. The transition from a condensed state to a fragmented state is clearly visible. At $A = 19$ the BMF solution does not reproduce the many-body results. The system is in a true many-body state, inaccessible to mean-field methods. At even higher barriers $A \geq 24$ the system fully fragments, and a mean-field description is applicable again. The quantities shown are dimensionless

condensate has now vanished completely in favor of a pattern characteristic of two separate condensates. The pattern agrees well with the structure predicted by Eq. 5.24.

Similar to our results on first-order correlations, discussed in Sect. 5.2, the fragmentation transition shows up in the two-particle momentum distribution, but not in the two-particle density. While this behavior is predictable in the limiting cases of a condensed and a fully fragmented state, it is necessary to solve the many-body problem to determine the limits of such mean-field approximations. Particularly the behavior in between the two mean-field limits is only accessible to many-body approaches.

We will now address the second-order coherence of the system. The first two columns of Fig. 5.6 show the diagonal of the normalized second-order correlation function $g^{(2)}(x_1, x_2, x_1, x_2)$ of the many-body (left) and the BMF (right) solutions. Note the scale! The Eqs. 5.17 and 5.26 of the analytical mean-field model of Sect. 5.1 predict very small correlations in the two-particle density of the condensed and the fragmented state. This is confirmed in the first column of Fig. 5.6. In the condensed regime at zero barrier height the effects of the depletion of the condensate on $g^{(2)}(x_1, x_2, x_1, x_2)$ are visible. Almost no two-particle density correlations are present. This is equally true in the case of a single condensate spread out over the two wells and also in the many-body regime. Above the fragmentation transition, the present computation of the many-body solution cannot describe effects on $g^{(2)}$ that are due to the depletion of the condensate. However, since depletion effects are negligible in the harmonic trap, we are reassured that they are also negligible in the fully fragmented regime, see Sect. 5.3.5. The BMF solution predicts almost identical two-body density correlations, see second column of Fig. 5.6.

On the basis of $g^{(2)}(x_1, x_2, x_1, x_2)$ alone, the many-body state *appears* to be second-order coherent at all barrier heights. A high degree of second-order coherence requires Eq. 2.30 to hold to a very good approximation for $p=1$ and $p=2$. This in turn requires the largest eigenvalues of the first- and second-order RDM to be $n_1^{(1)} \approx N$ and $n_1^{(2)} \approx N(N-1)$, respectively. We have already demonstrated in Sect. 5.3 that these conditions are only satisfied in the condensed regime. Therefore, it is obviously tempting, but wrong to conclude from $g^{(2)}(x_1, x_2, x_1, x_2) \approx 1$ that the system is second-order coherent. This misconception is due to the fact that $g^{(2)}(x_1, x_2, x_1, x_2)$ only samples a small part of the first- and second-order RDMs of the system.

So, how does the decrease of coherence manifest itself in second-order correlation functions? For second-order coherence to be present, at least approximately, also $g^{(2)}(k_1, k_2, k_1, k_2)$ has to be close to one. The third and fourth column of Fig. 5.6 show $g^{(2)}(k_1, k_2, k_1, k_2)$ of the many-body (left) and BMF (right) solution. At zero barrier height the system is indeed highly second-order coherent since only one natural orbital is significantly occupied. Not only $g^{(2)}(x_1, x_2, x_1, x_2)$, but also $g^{(2)}(k_1, k_2, k_1, k_2)$ is very close to one here. However, at $A = 13$ when the many-body state is still condensed, $g^{(2)}(k_1, k_2, k_1, k_2)$ starts to develop a structure.

When the barrier is raised to values above $A = 13$, the structure becomes more and more pronounced. In the many-body regime at $A = 19$, we find that $g^{(2)}(k_1, k_2, k_1, k_2)$ has a complicated behavior and deviates significantly from values close to one, thereby proving that strong correlations are present. Note that the interaction between the particles is weak and that the strong correlations are due to the transition from a single to a fragmented condensate. This transition is in turn induced by a change of the shape of external potential. Varying the shape of the external potential therefore provides a means to introduce strong correlations between the particles. The strongest correlations (black spots in the third panel of the third row of Fig. 5.6)

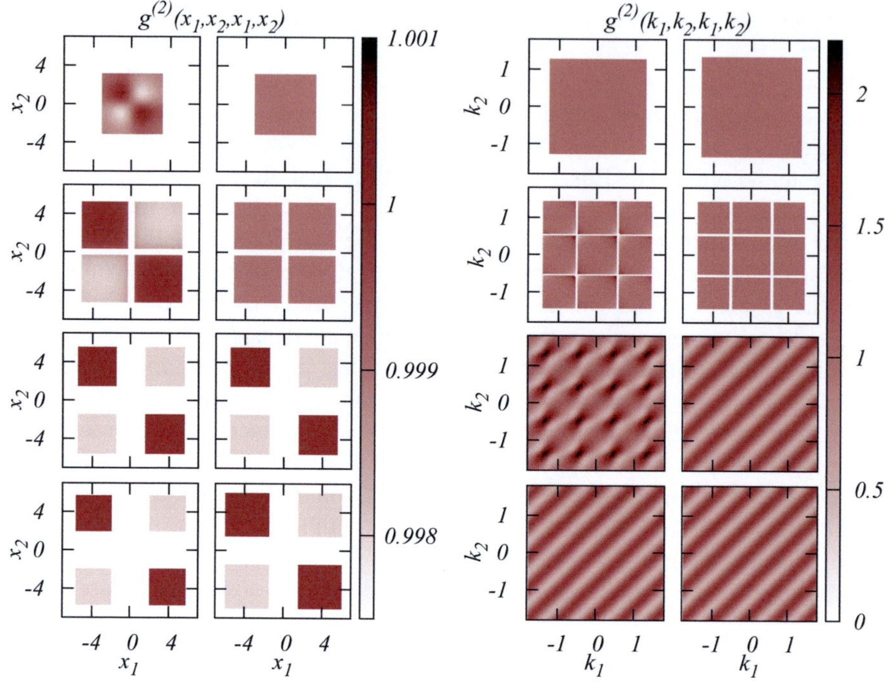

Fig. 5.6 Second-order coherence at different barrier heights. The first two columns (from *left* to *right*) show the diagonal of the normalized second-order correlation function in real space $g^{(2)}(x_1, x_2, x_1, x_2)$ of the many-body (*left*) and BMF (*right*) solution at barrier heights $A = 0$, 13, 19, 24, from top to bottom. $g^{(2)}(x_1, x_2, x_1, x_2)$ is very close to one at all barrier heights. Note the scale! The system seems to be second-order coherent and the results of the many-body and BMF solution agree well with each other. The third and fourth column depict the diagonal of the normalized second-order correlation function in momentum space $g^{(2)}(k_1, k_2, k_1, k_2)$ of the many-body (*left*) and BMF (*right*) solution at the same barrier heights. The fragmentation transition is clearly visible between $A = 13$ and $A = 24$. At $A = 19$ there are strong many-body correlations between the momenta (local maxima in *black color*) and $g^{(2)}(k_1, k_2, k_1, k_2)$ exhibits a complicated pattern, see text for more details. A mean-field description fails here. In the limit of high barriers, $A \geq 24$, the correlations of the many-body state become again describable by those of the BMF solution. The quantities shown are dimensionless

occur at those values where the two-body momentum distribution has local minima. At the values of k_1 and k_2, where the strongest correlations occur, the one-body and the two-body momentum distributions are *clearly* distinct from zero, see third panel in the middle column of Fig. 5.3 and the third panel in the third row of Fig. 5.5. Experiments that measure $g^{(2)}(k_1, k_2, k_1, k_2)$ in ultracold quantum gases have been carried out recently, see e.g. [44]. An experiment that measures $g^{(2)}(k_1, k_2, k_1, k_2)$ would find the strongest two-particle momentum correlations at intermediate barrier heights.

When the system becomes fully fragmented at barrier heights $A \geq 24$ the structure of $g^{(2)}(k_1, k_2, k_1, k_2)$ becomes more regular again. The amplitude of the correlations is smaller than in the many-body regime, and the correlations between different momenta are modulated by a single oscillatory structure. This structure can be well understood within the analytical mean-field model of Sect. 5.5.1. The oscillatory part of $g^{(2)}(k_1, k_2, k_1, k_2)$ is determined by the difference of the wave vectors multiplied by the distance between the wells, see Eq. 5.27. In contrast, the correlations in the many-body regime cannot be explained by analytical mean-field models.

Hence, we find that only in the condensed regime the system is second-order coherent despite the fact that $g^{(2)}(x_1, x_2, x_1, x_2) \approx 1$ at all barrier heights. This merely reflects the fact that $g^{(2)}(x_1, x_2, x_1, x_2)$ is only the diagonal of $g^{(2)}(x_1', x_2', x_1, x_2)$. On the other hand, $g^{(2)}(k_1, k_2, k_1, k_2)$ depends on all values of of $\rho^{(2)}(x_1, x_2|x_1', x_2')$ and provides complementary information about the coherence of the state. A description of second-order coherence in terms of $g^{(2)}(x_1, x_2, x_1, x_2)$ alone is therefore incomplete.

The corresponding results of the BMF solution agree well with those of the many-body solution as long as the system is not in the many-body regime at intermediate barrier heights. In the many-body regime the BMF result is inaccurate, but it anticipates the final form of $g^{(2)}(x_1, x_2, x_1, x_2)$ in the fragmented regime.

5.6 Conclusions

In this Chapter we have investigated first- and second-order correlations of trapped interacting bosons. For illustration purposes we have investigated the ground state of $N = 1{,}000$ weakly interacting bosons in a one-dimensional double-well trap geometry at various barrier heights on a many-body level. The interaction strength was such that according to the classification scheme in Sect. 2.11 the system should have been in the 1D Thomas-Fermi regime where many-body effects are highly suppressed and the ground state should have been of the Gross–Pitaevskii type at any barrier height. This was not the case. The work in this Chapter thereby proves that great care must be taken when exact results for infinite numbers of particles are taken over to finite systems. Obviously, $N = 1{,}000$ particles are not enough to justify the limit of an infinite number of particles.

We have obtained the many-body results by solving the many-body Schrödinger equation with the recently developed MCTDHB method. This allowed us to compute from first principles the natural orbitals and the natural geminals of a large interacting many-body system, together with their occupation numbers. To our knowledge this is the first computation of the natural geminals of an interacting many-body system of this size.

Depending on the height of the double-well barrier we found that there are three different parameter regimes. At low barriers the ground state is condensed and the many-body wave function is well approximated by a single permanent of the form

$|N, 0\rangle$. At high barriers the ground state becomes fully fragmented and can be well approximated by a single permanent of the form $|N/2, N/2\rangle$. At intermediate barrier heights, where the transition from a single to a fragmented condensate occurs, the ground state becomes a true many-body wave function to which many permanents contribute. We have demonstrated that the transition to a fragmented state results in the occupation of a natural geminal that contributes very little to the interaction energy. The overall energy of the system can be lowered by the occupation of such a geminal, and the ground state becomes fragmented.

We have shown how the transition from a condensed to a fully fragmented ground state manifests itself in the one- and two-particle momentum distributions. However, the transition is *not* captured by the one- and two-particle density distributions, not even in the many-body regime.

In order to determine the coherence of the state during the fragmentation transition, we have computed the first- and second-order normalized correlation functions $g^{(1)}(x_1', x_1)$, $g^{(2)}(x_1, x_2, x_1, x_2)$ and $g^{(2)}(k_1, k_2, k_1, k_2)$. In the condensed regime, a high degree of coherence is indeed present in the ground state wave function. First and second-order correlations were found to be negligible at the interaction strength and particle number chosen for our computation. However, with increasing barrier height correlations between the momenta of the particles build up. These correlations were found to be very strong in the many-body regime at intermediate barrier heights. The ground state at high barriers was found to be correlated, but not as strongly as the ground state at intermediate barrier heights.

While the transition from a virtually uncorrelated state to a correlated one is clearly visible in $g^{(1)}(x_1', x_1)$ and $g^{(2)}(k_1, k_2, k_1, k_2)$, the transition hardly shows up in $g^{(2)}(x_1, x_2, x_1, x_2)$. A description of second-order coherence in terms of $g^{(2)}(x_1, x_2, x_1, x_2)$ alone is, therefore, incomplete and can lead to wrong predictions.

For comparison we have computed results based on (i) the best approximation of the many-body wave function within mean-field theory, the BMF wave function, and (ii) the Gross–Pitaevskii solution. We found that the GP wave function is identical to the BMF solution up to some barrier height. However, once the true many-body solution starts to fragment the BMF wave function is no longer given by a GP type permanent $|N, 0\rangle$, but rather by a fragmented state of the form $|N/2, N/2\rangle$. In the true many-body regime neither the GP, nor the BMF solution provide an adequate approximation to the many-body wave function, and the predicted correlations are inaccurate.

While the GP mean-field is only accurate at low barrier heights, the BMF solution provides a very good approximation to the true many-body wave function at low *and* high barriers. We have shown that the GP mean-field predicts qualitatively wrong results at high barriers. The BMF only fails at intermediate barrier heights where the true many-body wave function becomes a superposition of many permanents. Such many-body effects can, by construction, not be captured by mean-field methods.

In the mean-field regimes at high and low barriers we have provided an analytical mean-field model that allows us to understand the general structure of the computed correlation functions.

Our work sheds new light on the first- and second-order correlation functions of interacting many-body systems. The variation of the shape of the trapping potential allows one to change the physics of the system from mean-field to strongly correlated many-body physics. Particularly, the many-body regime in between the condensed and the fully fragmented regimes was shown to be very rich and promises exciting results for experiments to come.

References

1. K. Sakmann, A.I. Streltsov, O. E. Alon, L. S. Cederbaum, Phys. Rev. A **78**, 023615 (2008)
2. M.H. Anderson, J.R. Ensher, M.R. Matthews, C.E. Cornell, E.A. Wieman, Science **269**, 198 (1995)
3. C.C. Bradley, C.A. Sackett, J.J. Tollett, R.G. Hulet, Phys. Rev. Lett. **75**, 1687 (1995)
4. K.B. Davis, M.-O. Mewes, M.R. Andrews, van N.J. Druten, D.S. Durfee, D.M. Kurn, W. Ketterle, Phys. Rev. Lett. **75**, 3969 (1995)
5. E.A. Burt, R.W. Ghrist, C.J. Myatt, M.J. Holland, E.A. Cornell, C.E. Wieman, Phys. Rev. Lett. **79**, 337 (1997)
6. L. Cacciapuoti, D. Hellweg, M. Kottke, T. Schulte, W. Ertmer, J.J. Arlt, K. Sengstock, L. Santos, M. Lewenstein, Phys. Rev. A **68**, 053612 (2003)
7. B.L. Tolra, K.M. O'Hara, J.H. Huckans, W.D. Phillips, S.L. Rolston, J.V. Porto, Phys. Rev. Lett. **92**, 190401 (2004)
8. S. Fölling, F. Gerbier, A. Widera, O. Mandel, T. Gericke, I. Bloch, Nature **434**, 481 (2005)
9. M. Schellekens, R. Hoppeler, A. Perrin, J.V. Gomes, D. Boiron, A. Aspect, C.I. Westbrook, Science **310**, 648 (2005)
10. A. Ottl, S. Ritter, M. Köhl, T. Esslinger, Phys. Rev. Lett. **95**, 090404 (2005)
11. L.E. Sadler, J.M. Higbie, S.R. Leslie, M. Vengalattore, D.M. Stamper-Kurn, Phys. Rev. Lett. **98**, 110401 (2007)
12. S. Ritter, A. Ottl, T. Donner, T. Bourdel, M. Köhl, T. Esslinger, Phys. Rev. Lett. **98**, 090402 (2007)
13. M. Naraschewski, R.J. Glauber, Phys. Rev. A **59**, 4595 (1999)
14. M. Holzmann, Y. Castin, Eur. Phys. J. D **7**, 425 (1999)
15. G.E. Astrakharchik, S. Giorgini, Phys. Rev. A **68**, 031602 (2003)
16. K.V. Kheruntsyan, D.M. Gangardt, P.D. Drummond, G.V. Shylapnikov, Phys. Rev. Lett. **91**, 040403 (2003)
17. R. Bach, Żewski K. Rza, Phys. Rev. A **70**, 063622 70 (2004)
18. J.V. Gomes, A. Perrin, M. Schellekens, D. Boiron, C.I. Westbrook, M. Belsley, Phys. Rev. A **74**, 053607 (2006)
19. R. Pezer, H. Buljan, Phys. Rev. Lett. **98**, 240403 (2007)
20. O. Penrose, L. Onsager, Phys. Rev. **104**, 576 (1956)
21. P. Nozières, D. Saint James, J. Phys. (Paris) **43**, 1133 (1982)
22. P. Nozières, in Bose- Einstein Condensation, edited by A. Griffin, D. W. Snoke, and S. Stringari (Cambridge University Press, Cambridge, England, 1996).
23. R.W. Spekkens, J.E. Sipe, Phys. Rev. A **59**, 3868 (1999)
24. L.S. Cederbaum, A.I. Streltsov, Phys. Lett. A **318**, 564 (2003)
25. A.I. Streltsov, L.S. Cederbaum, N. Moiseyev, Phys. Rev. A **70**, 053607 (2004)
26. O.E. Alon, L.S. Cederbaum, Phys. Rev. Lett. **95**, 140402 (2005)

27. O.E. Alon, A.I. Streltsov, L.S. Cederbaum, Phys. Lett. A **347**, 88 (2005)
28. A.I. Streltsov, O.E. Alon, L.S. Cederbaum, Phys. Rev. A **73**, 063626 (2006)
29. E.H. Lieb, R. Seiringer, Phys. Rev. Lett. **91**, 150401 (2003)
30. E.H. Lieb, R. Seiringer, J.P. Solovej, J. Yngvason, The Mathematics of the Bose Gas and its Condensation. (Birkhäuser, Verlag, 2000)
31. A.I. Streltsov, O.E. Alon, L.S. Cederbaum, Phys. Rev. Lett. **99**, 030402 (2007)
32. O.E. Alon, A.I. Streltsov, L.S. Cederbaum, Phys. Rev. A **77**, 033613 (2008)
33. O.E. Alon, A.I. Streltsov, L.S. Cederbaum, Phys. Rev. Lett. **95**, 030405 (2005)
34. M.J. Girardeau, Math. Phys. **1**, 516 (1960)
35. J.B. McGuire, J. Math. Phys. **5**, 622 (1964)
36. K. Sakmann, A.I. Streltsov, O.E. Alon, L.S. Cederbaum, Phys. Rev. A **72**, 033613 (2005)
37. E.H. Lieb, W. Liniger, Phys. Rev. **130**, 1605 (1963)
38. M.D. Girardeau, E.M. Wright, Phys. Rev. Lett. **84**, 5691 (2000)
39. V.I. Yukalov, M.D. Girardeau, Laser Phys. Lett. **2**, 375 (2005)
40. M.D. Girardeau, Phys. Rev. Lett. **97**, 100402 (2006)
41. A.G. Sykes, P.D. Drummond, M.J. Davis, Phys. Rev. A **76**, 063620 (2007)
42. H. Buljan, R. Pezer, T. Gasenzer, Phys. Rev. Lett. **100**, 080406 (2008)
43. L. Pitaevskii, S. Stringari, Phys. Rev. Lett. **83**, 4237 (1999)
44. A. Perrin, H. Chang, V. Krachmalnicoff, M. Schellekens, D. Boiron, A. Aspect, C.I. Westbrook, Phys. Rev. Lett. **99**, 150405 (2007)

Chapter 6
Exact Quantum Dynamics of a Bosonic Josephson Junction

The quantum dynamics of a one-dimensional bosonic Josephson junction is studied by solving the time-dependent many-boson Schrödinger equation numerically exactly. Already for weak interparticle interactions and on short time scales, the commonly-employed mean-field and many-body methods are found to deviate substantially from the exact dynamics. The system exhibits rich many-body dynamics like enhanced tunneling and a novel equilibration phenomenon of the junction depending on the interaction, attributed to a quick loss of coherence. Most results of this chapter were first published in Ref. [1]. Here we have added a detailed discussion of the fragmentation of the condensate and further studies in the self-trapping regime.

6.1 Introduction

Recent experiments on interacting Bose–Einstein condensates in double-well traps have led to some of the most exciting results in quantum physics, including matter-wave interferometry [2, 3], squeezing and entanglement [4, 5, 6] as well as work on high-precision sensors [7]. Particular attention has been paid to tunneling phenomena of interacting Bose–Einstein condensates in double-wells, which in this context are referred to as bosonic Josephson junctions. Explicitly, Josephson oscillations and self-trapping (suppression of tunneling) with Bose–Einstein condensates have been predicted [8, 9] and recently realized in experiments [10, 11], drawing intensive interest, see, e.g., [1, 8–22] and references therein. Nearly all works available in the literature on Josephson junctions are for repulsive interaction. In the following we therefore assume repulsive interaction.

In Chap. 5 we have investigated the correlation functions and the coherence of the ground state of a trapped condensate. Here we study the dynamics of a trapped condensate. Until recently the dynamics of bosonic Josephson junctions had been studied exclusively on the basis of Gross–Pitaevskii theory, discrete Gross–Pitaevskii models and on the basis of the Bose–Hubbard model, but not on the basis of the

K. Sakmann, *Many-Body Schrödinger Dynamics of Bose–Einstein Condensates*, Springer Theses, DOI: 10.1007/978-3-642-22866-7_6,

many-body Schrödinger equation. Here we fill this gap by providing the first numerically exact results in literature on the many-body quantum dynamics of a 1D bosonic Josephson junction [1]. This is made possible by a breakthrough in the solution of the time-dependent many-boson Schrödinger equation. We use the exact solution to check the current understanding of bosonic Josephson junctions—commonly described by the popular Gross–Pitaevskii mean-field theory and the Bose–Hubbard many-body model—and to find novel phenomena. The results of the Gross–Pitaevskii and Bose–Hubbard theories are found to deviate substantially from the full many-body solution, already for weak interactions and on short time scales. In particular, the well-known self-trapping effect is greatly reduced. We attribute these findings to a quick loss of the junction's coherence not captured by the common methods. For stronger interactions and on longer time scales, we find a novel equilibration dynamics in which the density and other observables of the junction tend towards stationary values. We show that the dynamics of bosonic Josephson junctions is much richer than what is currently known.

6.2 Theories for Bosonic Josephson Junctions

In the following we assume that the bosonic Josephson junction consists of N interacting bosons trapped in an external, symmetric double-well potential.

6.2.1 *Exact Many-Body Schrödinger Dynamics*

To compute the exact time evolution of a 1D bosonic Josephson junction, we solve the time-dependent many-boson Schrödinger equation using the MCTDHB method, as explained in Sect. 3.1 and Refs. [23, 24]. To be precise, we solve the equation

$$i\frac{\partial}{\partial t}\left|\Psi(t)\right\rangle = \hat{H}\left|\Psi(t)\right\rangle \tag{6.1}$$

numerically, where $\hat{H}$ is the *full* many-body Hamiltonian given in Eq. 2.8 and the *ansatz* for the many-boson wave function $|\Psi(t)\rangle$ is of the most general form (2.15). By successively increasing the number of time-dependent orbitals M in the ansatz wave function (2.15) we obtain convergence and thereby numerically exact results for a large number of particles. The present results rely on a novel mapping of the many-boson configuration space in combination with a parallel implementation of MCTDHB, allowing the efficient handling of millions of time-dependent, optimized permanents [25].

6.2.2 *Bose–Hubbard Many-Body Dynamics*

A popular approximative method for the description of bosonic Josephson junctions is the Bose–Hubbard many-body model restricted to two sites [8, 13]. In Sect. 4.3 it was shown under what assumptions the Bose–Hubbard Hamiltonian (4.14) can be obtained from the full Hamiltonian (2.8), and what assumptions are made about the *ansatz* for the Bose–Hubbard wave function (4.10). We refer the reader to Sect. 4.3 for further details on the Bose–Hubbard model. We will compare predictions of the Bose–Hubbard model with the results of the exact solution of the time-dependent many-body Schrödinger equation.

6.2.3 *Gross–Pitaevskii and Two-Mode Gross–Pitaevskii Mean-Field Dynamics*

Gross–Pitaevskii theory is a popular approximative method in the theory of Bose–Einstein condensates and it was shown in Sect. 3.2 that Gross–Pitaevskii theory is a special case of the MCTDHB method, namely the case, where precisely one orbital is used. The *ansatz* for the many-body wave function, Eq. 2.15, then consists of a single permanent in which all bosons reside in one time-dependent orbital. For more details on Gross–Pitaevskii theory see Sect. 3.2. We will compare the predictions of Gross–Pitaevskii theory with those of the exact solution of the time-dependent many-body Schrödinger equation. In the context of bosonic Josephson junctions an approximation to Gross–Pitaevskii theory, known as the two-mode Gross–Pitaevskii model is actually more popular than Gross–Pitaevskii theory [8, 9]. We have reviewed the two-mode Gross–Pitaevskii model earlier and refer the reader to Sect. 4.2 for further details.

6.3 Observables of the Bosonic Josephson Junction

Having computed the many-boson wave function $|\Psi(t)\rangle$, we focus on the evolution of the following quantities to analyze the dynamics of the Josephson junction. The eigenvalues $n_i^{(1)}$ of the first order RDM, $\rho^{(1)}(x|x';t)$, Eq. 2.33, determine the extent to which the system is condensed or fragmented. They are known as natural orbital occupations and their role in Bose–Einstein condensation has been discussed in Sect. 2.10. As is common in the analysis of bosonic Josephson junctions, the "survival probability" of the system in, e.g., the left well, is obtained by integrating the density $\rho(x;t) \equiv \rho^{(1)}(x|x'=x;t)$ over the left well,

$$p_L(t) \equiv \frac{1}{N}\int_{-\infty}^{0} dx\rho(x;t). \tag{6.2}$$

In the exposition of two-mode Gross–Pitaevskii theory in Sect. 4.2 we have touched upon the point of self-trapping. In order to quantify self-trapping it will be useful to consider the time-averaged survival probability

$$\bar{p}_L(T) = \frac{1}{T}\int_0^T dt p_L(t). \tag{6.3}$$

As was shown in Ref. [26] the analytical solution of the two-mode Gross–Pitaevskii model discussed in Sect. 4.2 can remain trapped for infinite times, $\bar{p}_L(T \to \infty) \neq 0.5$. Such solutions cannot exist on the many-body level, since the potential is symmetric and therefore all eigenstates are parity eigenfunctions. For finite barrier height they are non-degenerate and hence the density always tunnels through the barrier. The question is only how long does it take. On the many-body level self-trapping can therefore only exist for a finite time and $\bar{p}_L(T \to \infty) = 0.5$. Finally, we use the normalized first-order correlation function

$$g^{(1)}(x', x; t) = \frac{\rho^{(1)}(x|x'; t)}{\sqrt{\rho(x, t)\rho(x', t)}} \tag{6.4}$$

to quantify the system's degree of spatial coherence [27, 28]. Normalized correlation functions were introduced in Sect. 2.8 and their signatures in experiments have been summarized in Sect. 2.9.

6.4 Details of the Bosonic Josephson Junction

We now turn to the details of the 1D bosonic Josephson junction considered in this work. The Hamiltonian used in this chapter is of the form given in Eq. 2.6.

We employ a contact potential $W(x - x') = \lambda_0\delta(x - x')$ for the interparticle interaction potential. To parameterize the interaction strength we use the parameter $\lambda = \lambda_0(N - 1)$, which appears naturally in the full many-body treatment and also in Gross–Pitaevskii theory. We quote the corresponding values of Λ, the interaction parameter in the two-mode Gross–Pitaevskii model defined in Eq. 4.8, and the values of U/J, which parameterizes the interaction strength within the Bose–Hubbard model.

The symmetric double-well potential *V(x)* is generated by connecting two harmonic potentials $V_\pm(x) = \frac{1}{2}(x \pm 2)^2$ with a cubic spline in the region $|x| \leq 0.5$. The lowest four single-particle energy levels $e_1 = 0.473$, $e_2 = 0.518$, $e_3 = 1.352$ and $e_4 = 1.611$ of *V(x)* are lower than the barrier height $V(0) = 1.667$. The trap parameters defined in Eq. 4.3 then take on the following values. The width of the lowest band is $2J = e_2 - e_1 = 0.045$ which is much smaller than the band gap energy, Eq. 4.4, $e_{gap} = e_3 - e_2 = 0.834$. With this choice of units the on-site one-body energy is $\epsilon = h_{LL} = h_{RR} = 0.495$. For noninteracting particles localized in one of the orbitals $\phi_{L,R}$ the tunneling oscillation period is $t_{Rabi} = \pi/J = 140.66$.

As we are working in 1D here λ_0 is determined by the scattering length a_s and the transverse confinement $\omega_\perp$ [29]. We now give some realistic experimental parameters for the cases considered below. As an example we choose $L = 1\mu$m and ^{87}Rb as a boson. Note that other realistic choices can be made. The unit of energy $\frac{\hbar^2}{mL^2}$

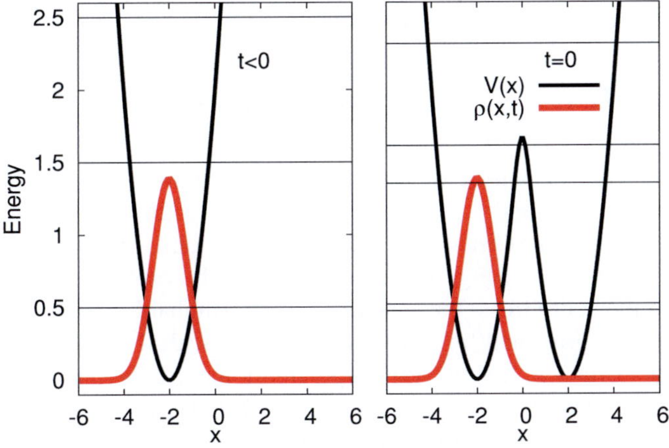

Fig. 6.1 Preparation procedure. Shown is the density (*red line*) of the many-body ground state in one trap (*left*) and the subsequent switch to a different potential in which the system evolves in time (*right*). The trapping potential (*thick black line*) changes from harmonic (*left*) to a symmetric double well (*right*). The parameter values shown for the system shown here are $M = 8$, $N = 20$, $\lambda = \lambda_0(N-1) = 0.152$. Also shown are the single-particle energy levels (*thin black lines*) in the respective trapping potentials. The lowest four single-particle energy levels $e_1 = 0.473$, $e_2 = 0.518$, $e_3 = 1.352$ and $e_4 = 1.611$ of the double-well potential *V(x)* (*right*) are lower than the barrier height $V(0) = 1.667$

then corresponds to $(2\pi\hbar)116$ Hz and the potential *V(x)* has the following characteristics: barrier height *V*(0): $(2\pi\hbar)193.9$ Hz, distance between the minima: 4μm, width (FWHM): 1.418μm, $t_{Rabi} = 192.4$ ms.

To realize the cases considered in this work a_s and $\omega_\perp$ can be chosen as follows. For $\lambda = 0.152$, $N = 20$ (100): $a_s = 1.28$ (0.246) nm, $\frac{\omega_\perp}{2\pi} = 363.4$ Hz. For $\lambda = 0.245$, $N = 20$ (100): $a_s = 2.06$ (0.396) nm, $\frac{\omega_\perp}{2\pi}$ =363.4 Hz. For $\lambda = 4.9$, $N = 10$ (100): $a_s = 5.31$nm, $\frac{\omega_\perp}{2\pi}$ =5962 (542.0) Hz.

6.5 Preparation and Propagation of the Many-Boson Wave Function

In all our computations *N* bosons are prepared at $t = 0$ in the exact many-body ground state of the potential $V_+(x)$ and then propagated in the potential *V(x)* as depicted schematically in Fig. 6.1. Within the Bose–Hubbard framework this procedure amounts to starting from the state in which all bosons occupy the orbital ϕ_L, i.e. from the state $|n_L, n_R\rangle = |N, 0\rangle$. As a guideline for the interaction strength we note that within the framework of the two-mode Gross–Pitaevskii model, such initial states are predicted to remain self-trapped if $\Lambda > \Lambda_c = 2$ [8, 9]. We will consider interaction strengths below, in the vicinity of and above Λ_c.

6.6 Results for Weak Interactions

6.6.1 Below the Self-Trapping Transition Point $\Lambda < \Lambda_c$

We begin our studies with a weak interaction strength $\lambda = 0.152$, leading to $U/J = 0.140$ (0.027) and $\Lambda = 1.40$ (1.35) for $N = 20$ (100) bosons, which is well below the two-mode Gross–Pitaevskii transition point for self-trapping $\Lambda_c = 2$. In the upper two panels of Fig. 6.2 the full many-body results for $p_L(t)$ are shown together with those of Gross–Pitaevskii and Bose–Hubbard theory. The full many-body dynamics is governed by three different time scales. On a time scale of the order of a Rabi cycle, $p_L(t)$ performs large-amplitude oscillations about $p_L = 0.5$, the long time average of $p_L(t)$. The amplitude of these oscillations is damped out on a time scale of a few Rabi cycles and marks the beginning of a collapse and revival (not shown) sequence [8], which is also found on the full many-body level. On top of these slow large-amplitude oscillations, a higher frequency with a small amplitude can be seen. In a single-particle picture these high frequency oscillations can be related to contributions from higher excited states in the initial wave function. However, a single particle picture fails to describe the dynamics, as we shall now show.

While the initial wave function $|\Psi(t = 0)\rangle$ is practically fully condensed—the fragmentation of the system is less than 10^{-4} (10^{-5}) for $N = 20$ (100) bosons—the propagated wave function $|\Psi(t)\rangle$ quickly becomes fragmented. In the upper two panels of Fig. 6.3 the fragmentation of the full many-body results are shown together with those of Gross–Pitaevskii and Bose–Hubbard theory. The fragmentation increases to about 33% (26%) at $t = 3t_{Rabi}$ for $N = 20$ (100) particles. If we evaluate the Lieb-Liniger parameter γ, as defined in Eq. 2.43 we find $\gamma N^2 = 0.41$ (0.39). According to the classification scheme in Sect. 2.11 the interactions are very weak and the system is at the border of the ideal gas regime to the 1D Gross–Pitaevskii regime. We remind the reader that the classification scheme in Sect. 2.11 is motivated by rigorous mathematical results for the ground states of trapped Bose-Einstein condensates, in the limit $N \to \infty$ at constant λ. Based on these results one may speculate that also the exact, finite N dynamics approaches the Gross–Pitaevskii result if N is increased at constant λ. Based on the available exact data there is no such trend. For both particle numbers we find strong fragmentation after short times, making a many-body treatment indispensable, already at this weak interaction strength. However, we note that the exact solution for 100 bosons fragments slightly slower here than that for 20 bosons.

The respective Gross–Pitaevskii results for $p_L(t)$ oscillate back and forth at a frequency close to the Rabi frequency and resemble the full many-body dynamics only on a time scale *shorter* than half a Rabi cycle. The poor quality of the Gross–Pitaevskii mean-field approximation is, of course, due to the fact that the exact wave function starts to fragment while the Gross–Pitaevskii dynamics remains condensed by construction.

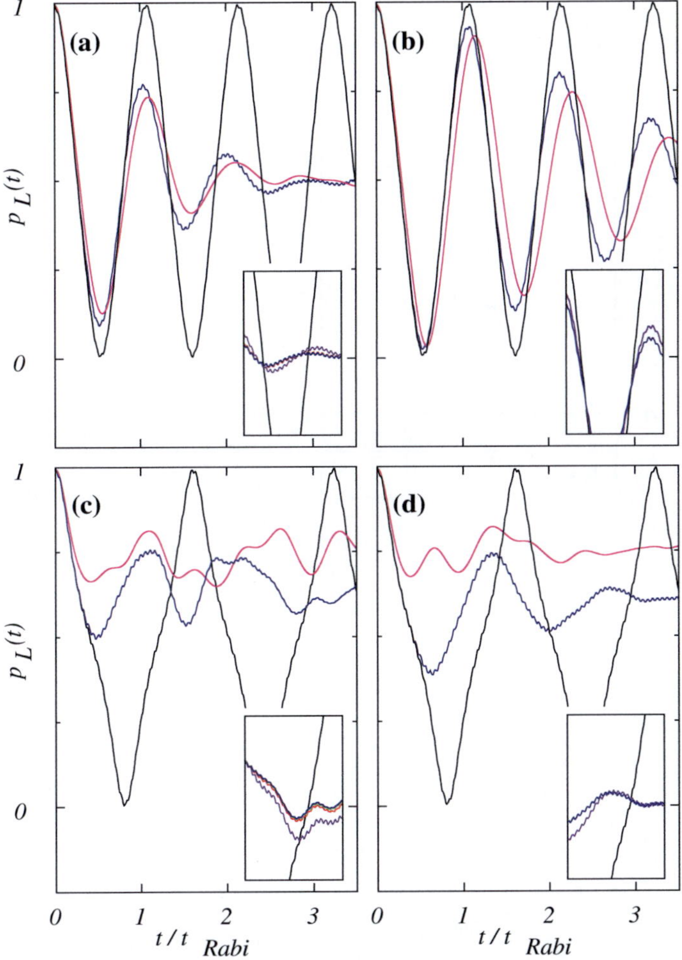

Fig. 6.2 Full quantum dynamics of a 1D bosonic Josephson junction below ($\Lambda < \Lambda_c$) and above ($\Lambda > \Lambda_c$) the transition to self-trapping as defined by the two-mode Gross–Pitaevskii theory. Shown is the full many-body result (*solid blue lines*) for the probability of finding a boson in the left well, $p_L(t)$. For comparison, the respective Gross–Pitaevskii (*solid black lines*) and Bose–Hubbard (*solid magenta lines*) results are shown as well. The parameter values are: **a** $N = 20,\ \lambda = 0.152$ and **b** $N = 100,\ \lambda = 0.152$ ($\Lambda < \Lambda_c$), **c** $N = 20,\ \lambda = 0.245$ and **d** $N = 100,\ \lambda = 0.245$ ($\Lambda > \Lambda_c$). The Gross–Pitaevskii and Bose–Hubbard results are found to deviate from the full many-body results already after short times. The insets show the convergence of the full many-body results. **a**,**c**: $M = 2$ (*solid purple line*), $M = 4$ (*solid red line*), $M = 6$ (*solid green line*), $M = 8$ (*solid blue line*). The $M = 2$ results are seen to deviate slightly from the converged results for $M \geq 4$. **b**,**d**: The results for $M = 2$ (*solid purple line*) and $M = 4$ (*solid blue line*) are shown. All quantities shown are dimensionless.

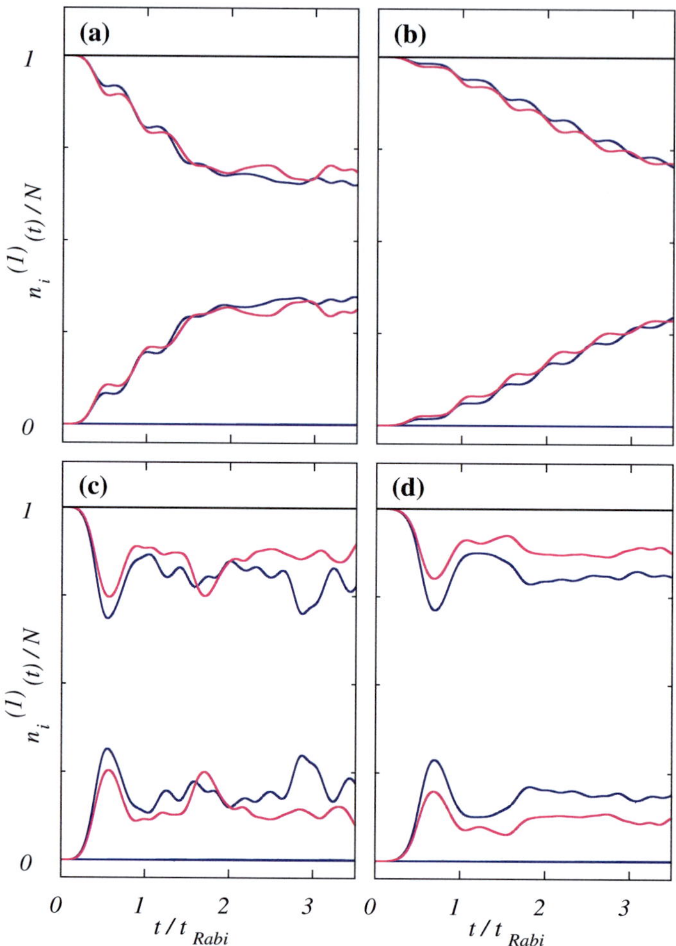

Fig. 6.3 Natural orbital occupations $n_i^{(1)}(t)$ for the same parameter values as in Fig. 6.2. Shown is the full many-body result (*solid blue lines*) together with the respective Gross–Pitaevskii (*solid black lines*) and Bose–Hubbard (*solid magenta lines*) results. The full many-body result is seen to fragment quickly in all cases shown. While the Bose–Hubbard results follow qualitatively the full many-body results for $\lambda = 0.152$ in (a) and (b), this is not the case for slightly stronger interaction at $\lambda = 0.245$ shown in (c) and (d). In all cases shown only two natural orbitals of the full many-body result are macroscopically occupied. The Gross–Pitaevskii result has only one natural orbital and is therefore restricted to a fully condensed system at all times. All quantities shown are dimensionless.

The Bose–Hubbard result for $p_L(t)$ reproduces many features of the exact solution at this interaction strength for both $N = 20$ and $N = 100$ particles. The large-amplitude oscillations collapse over a period of a few Rabi cycles and revive at a later stage (not shown). Also the Bose–Hubbard solution quickly becomes fragmented, starting from the left localized state, which is totally condensed. The fragmentation of the Bose–Hubbard wave function for $N = 20$ (100) particles at $t = 3t_{Rabi}$ is essentially the same as the respective value of the exact solution. However, differences between the exact and the Bose–Hubbard result are visible even on time scales less than half a Rabi cycle. Not only are the amplitudes obviously different, but also the frequencies contained in $p_L(t)$. Furthermore, the Bose–Hubbard solutions do not exhibit a high frequency oscillation on top of the slow large-amplitude oscillations; a difference which is related to the fact that the Bose–Hubbard orbitals are *time-independent* and thus, not determined variationally at each point in time. Note that $p_L(t)$ is a quantity in which *all* spatial degrees of freedom have been integrated out. Visible differences in $p_L(t)$ imply that it is not only the densities $\rho(x, t)$ which must differ, but also *all* correlation functions.

The insets of Fig. 6.2a, b demonstrate the convergence of the many-body dynamics results. In particular and somewhat unexpectedly, the number of *time-dependent* orbitals needed to describe the bosonic Josephson junction dynamics quantitatively is $M = 4$, even below the transition point for self-trapping. These orbitals are determined variationally *at each point in time*, implying that any method using time-independent orbitals will need substantially more orbitals to achieve the same accuracy.

6.6.2 Around the Self-Trapping Transition Point $\Lambda \approx \Lambda_c$

One of the central phenomena often discussed in the context of bosonic Josephson junctions is the celebrated transition to self-trapping [8–11]. In what follows we would like to study the dynamics of a bosonic Josephson junction from the full many-body perspective at an interaction strength where self-trapping is supposed to occur.

The interaction strength is taken to be $\lambda = 0.245$, leading to $U/J = 0.226$ (0.043) and $\Lambda = 2.26$ (2.17) for $N = 20$ (100) particles. Hence, the system is just above the critical value for self-trapping $\Lambda_c = 2$ [8, 9]. The results for $N = 20$ and $N = 100$ are collected in Fig. 6.2c, d. We find that the full many-body solutions exhibit indeed some self-trapping on the time scale shown, with $\bar{p}_L(3t_{Rabi}) = 0.66$ (60). The fragmentation of the condensate for $N = 20$ (100) bosons increases from initially less than 10^{-4} (10^{-5}) to about 28% (18%) after three Rabi cycles, as can be seen in the lower two panels of Fig. 6.3.

For the Lieb-Liniger parameter we find $\gamma N^2 = 0.66$ (0.64) here. Again, according to Sect. 2.11 the interactions are very weak and the system is at the border of the ideal gas regime to the 1D Gross–Pitaevskii regime. Nevertheless, Gross–Pitaevskii theory is—as before—inapplicable, even on time scales shorter than $t_{Rabi}/2$. Note also that the Gross–Pitaevskii result does not show any sign of self-trapping at this interaction strength, despite the fact that $\Lambda > \Lambda_c$. Interestingly, after three Rabi cycles the many-body results now fragment *less* than for weaker interaction. In contrast to the situation at weaker interaction both many-body results for $N = 20$ and $N = 100$ bosons seem to fragment similarly quickly for this λ. Based on these results there is no indication that the dynamics becomes less fragmented for larger numbers of particles at constant λ.

The Bose–Hubbard results deviate from the true dynamics even earlier. They greatly overestimate the self-trapping and coherence of the condensate. According to the Bose–Hubbard model the condensate would only be 13% (11%) fragmented for $N = 20$ (100) at $t = 3t_{Rabi}$, which is not the case. Similarly, we find $\bar{p}_L(3t_{Rabi}) = 0.74\,(0.76)$ for the Bose–Hubbard results, which is considerably larger than the respective results of the exact solution. This trend also continues for stronger interactions, see below.

Let us briefly discuss the applicability of Gross–Pitaevskii theory and the Bose–Hubbard model to the cases considered above. We have already evaluated the Lieb-Liniger parameter γ for the cases above and found that all cases discussed so far lie deeply in the regime of applicability of Gross–Pitaevskii theory according to the classification scheme given in Refs. [30, 31] and Sect. 2.11. Gross–Pitaevskii theory failed in all cases after short times. We have thereby clearly shown a failure of Gross–Pitaevskii theory within its range of expected validity.

According to the criterion (Eq. 4.17) discussed in Sect. 4.3 the Bose–Hubbard model is expected to be valid when the chemical potential μ is well below the band gap e_{gap} and the initial state lies within the first band [8]. These conditions are well satisfied. We find $\mu/e_{gap} \approx \frac{1}{14}$ and $\frac{1}{9}$ for the cases shown in Fig. 6.1a,b and Fig. 6.1c, d, respectively and independent of whether we use the full many-body solution or the Bose–Hubbard model to evaluate μ. The chemical potential is here computed as the energy difference of the initial state of $N+1$ and N particles at the same λ_0 *and* ϵ as defined in Eq. 4.3 is taken as the origin of energy.

The overlap integral of the initial states' first natural orbital with the left Bose–Hubbard orbital is 0.9991(!) in all cases; the initial states are therefore essentially given by the Bose–Hubbard state $|N, 0\rangle$. The results do not depend significantly on this tiny difference. We have thereby also shown a clear failure of the Bose–Hubbard model within its range of expected validity.

6.7 Results for Stronger Interactions

6.7.1 Self-Trapping at $\Lambda \approx 10\Lambda_c$

It was shown in Sect. 6.6.2 that the self-trapping effect of the full many-body solution increased with increasing interaction strength. Gross–Pitaevskii theory and the Bose–Hubbard model were shown to deviate substantially from the exact results at weak interaction strengths. However, as a trend we found that by *increasing* the interaction strength the system fragmented *less* over same periods of time. Should this trend continue for stronger interactions Gross–Pitaevskii theory would have to become *more* accurate with *increasing* interaction strength, a rather unexpected situation.

In order to investigate this more carefully, we consider $N = 20$ bosons at an interaction strength $\lambda = 1.96$, leading to $U/J = 1.81$ and $\Lambda = 18.1$. Hence, the system is high above the critical value for self-trapping $\Lambda_c = 2$ [8, 9]. Here, we find $\gamma N^2 \approx 6.2$ and according to Sect. 2.11 Gross–Pitaevskii theory should be applicable, since the system is at the border between the 1D Gross–Pitaevskii and the 1D Thomas-Fermi regime. Interestingly, at the same time the criterion (4.17) gives $\mu/e_{gap} \approx 1/3$ for the full many-body result and $\mu/e_{gap} \approx 1$ for the Bose–Hubbard model, which implies that based on the criterion (4.17) the Bose–Hubbard many-body model is not expected to be valid, but Gross–Pitaevskii mean-field theory *is*. There is clearly a discrepancy here between the criteria for Gross–Pitaevskii theory and the Bose–Hubbard model.

In the top panel of Fig. 6.4. the many-body result for $p_L(t)$ is shown together with the respective results of Gross–Pitaevskii theory and the Bose–Hubbard model. Note the change of scale! It is clearly seen that the self-trapping effect is much more pronounced on the time scale shown. The oscillations now have a much higher frequency and the Gross–Pitaevskii result follows the full many-body result for about half a Rabi cycle before it deviates again. Both the full many-body and the Gross–Pitaevskii result are self-trapped on the time scale shown with high frequency oscillations of $p_L(t)$ about the same value. More precisely, we find $\bar{p}_L(2t_{Rabi}) = 0.93$ for the full many-body and the Gross–Pitaevskii result. The Bose–Hubbard result does not follow the full many-body result at any time. It is almost completely trapped, $\bar{p}_L(2t_{Rabi}) = 0.99$, and oscillates sinusoidally with a very small amplitude. These sinusoidal oscillations can be understood by approximating the Bose–Hubbard Hamiltonian by a two-level system in the limit of strong interactions [32]. Obviously, the true dynamics is very different from the Bose–Hubbard dynamics.

The lower panel of Fig. 6.4 reveals that the fragmentation of the full many-body result is indeed much smaller here, than for the weaker interacting cases discussed in Sect. 6.6. This explains the relatively good performance of Gross–Pitaevskii theory. The Bose–Hubbard model, just like Gross–Pitaevskii theory does not fragment strongly at this interaction strength and describes a nearly condensed system. Here we find the peculiar situation where the Gross–Pitaevskii mean-field theory is more accurate than the Bose–Hubbard many-body model. This clearly demonstrates the importance of time-dependent, variationally determined orbitals: as long as the system remains largely condensed a mean-field theory using only one time-dependent orbital can give more accurate results than a many-body model employing two time-independent orbitals. Nevertheless, even though Gross–Pitaevskii theory gives qualitatively similar results we note that $M = 8$ orbitals were needed to obtain convergence for the results shown in Fig. 6.4. The full many-body dynamics remains complex, even though only one natural orbital is macroscopically occupied.

The following general statement about the relationship between self-trapping and coherence can be inferred from our full many-body results: The longer the system stays coherent the more self-trapping is present. We find this statement to be true at all interaction strengths and all particle numbers considered in this work.

As discussed in Sect. 4.2 self-trapping was first predicted based on the two-mode Gross–Pitaevskii model. It was also discussed earlier in this chapter that self-trapping

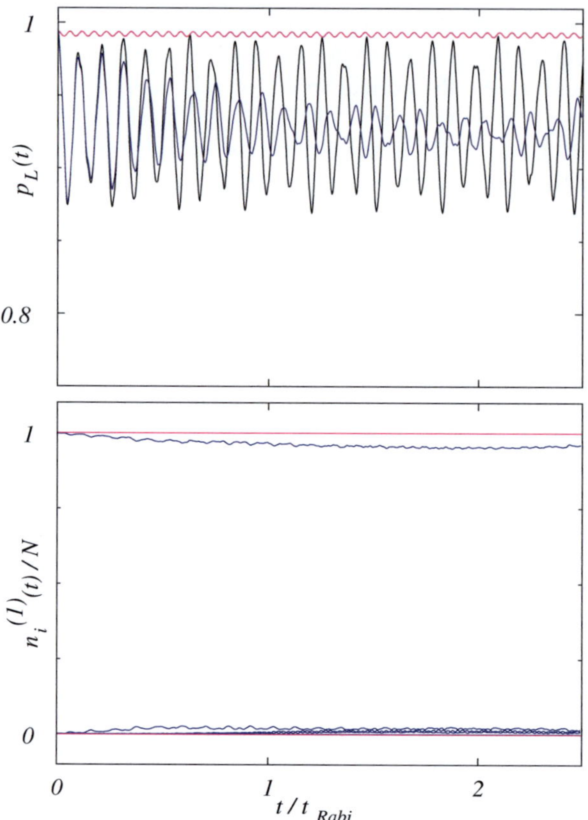

Fig. 6.4 Full quantum dynamics of a 1D bosonic Josephson junction at $\Lambda = 18.1$. *Top* as in Fig. 6.2 the full many-body result (*blue line*) for the probability of finding a boson in the left well, $p_L(t)$, is shown. The full many-body solution is self-trapped on the time scale shown with $\bar{p}_L(2t_{Rabi}) = 0.93$. Note the scale. *Bottom* as in Fig. 6.3 the natural orbital occupations are shown (*lines*). For comparison, the respective Gross–Pitaevskii (*solid black lines*) and Bose–Hubbard (*solid magenta lines*) results are shown as well. The parameter values are $N = 20$, $\lambda = 1.96$, $M = 8$. Since the full many-body solution shows little fragmentation the Gross–Pitaevskii result is close to the full many-body one. The Bose–Hubbard result differs qualitatively. All quantities shown are dimensionless.

in real physical systems can only exist for finite times because the density always tunnels from one well to the other. The question is only how long does it take. Predictions about the time scale on which self-trapping is lost have been published in the literature based on the Bose–Hubbard model, see e.g. Ref.[18], where cases up to $\Lambda = 500$ were considered. It can clearly be seen from results shown in Figs. 6.2 and 6.4 that the Bose–Hubbard model cannot be used to predict such quantities even if $\Lambda \approx \Lambda_c = 2$.

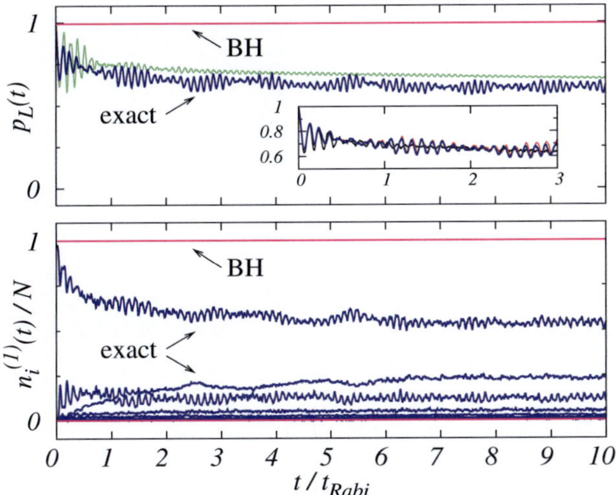

Fig. 6.5 Emergence of equilibration of the density at interaction strength $\lambda = 4.9$. *Top* same as Fig. 6.2, but for $N = 10$ (*solid blue line*) and $N = 100$ (*solid green line*). The respective Bose–Hubbard (*solid magenta lines*) results are on top of each other. In contrast to the Bose–Hubbard dynamics which is completely self-trapped, the full many-body dynamics is not. $p_L(t)$ tends towards its long-time average $p_L = 0.5$. For $N = 100$ particles $M = 4$ orbitals were used. The inset shows the convergence of the full many-body solution for $N = 10$ bosons: $M = 4$ (*solid black line*), $M = 10$ (*solid blue line*), $M = 12$ (*solid red line*). The $M = 4$ result follows the trend of the converged $M = 12$ result. *Bottom* corresponding natural orbital occupations for $N = 10$ bosons. The system becomes fragmented and roughly four natural orbitals are macroscopically occupied. All quantities are dimensionless.

6.7.2 Equilibration at $\Lambda < 25\Lambda_c$

Let us increase the interaction strength even further to $\lambda = 4.9$, which is even higher above the self-trapping transition point. This leads to $U/J = 9.55$ (0.869) and $\Lambda = 47.8$ (43.4) for $N = 10$ (100) bosons. Note that we now use ten instead of twenty bosons to demonstrate convergence. The energy per particle of the full many-body wave function is now $E/N = 1.22$ (1.28) for $N = 10$ (100) bosons, which is still below the barrier height $V(0) = 1.667$. We then find $\gamma N^2 = 9.7$ (17.7) and according to Sect. 2.11 Gross–Pitaevskii theory should be valid since the system is in the 1D Thomas-Fermi regime. At the same time the applicability criterion for the Bose–Hubbard model, Eq. 4.16, is not fulfilled since $\mu/e_{gap} = 1.6$ (1.7) if the full many-body solution is used to compute μ. Although here we are at least outside the range of expected validity of the Bose–Hubbard model, it is interesting to see what Gross–Pitaevskii theory and the Bose–Hubbard model fail to describe. Fig. 6.5(*top*) shows the full many-body results for $N = 10$ and $N = 100$ bosons together with those of the Bose–Hubbard model. The two Bose–Hubbard results lie on top of each other. In complete contrast to the Bose–Hubbard dynamics, for which $p_L(t)$ remains trapped in the left well, the full many-body dynamics shows no self-trapping. Instead,

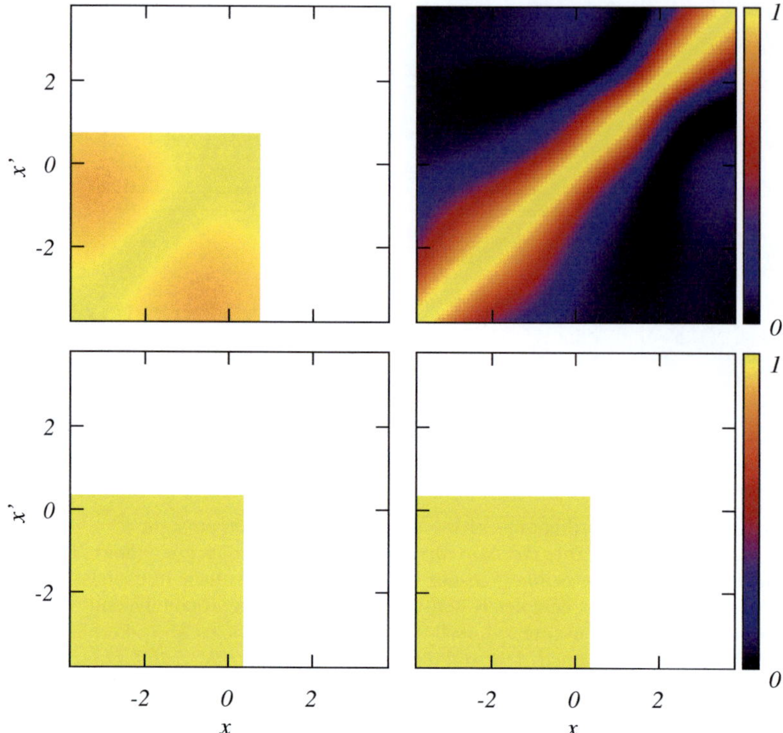

Fig. 6.6 Dynamics of the first order correlation function for $\lambda = 4.9$ at which the equilibration phenomenon of Fig. 6.5 occurs. Shown is $|g^{(1)}(x', x; t)|^2$ of $N = 10$ bosons at different times. *Top left* full many-body result at $t = 0$. The initial state exhibits coherence over the entire extent of the system. *Top right* full many-body result at $t = 10t_{Rabi}$. The coherence is lost. The system is incoherent even on short length scales. *Bottom left* Bose–Hubbard result at $t = 0$. *Bottom right* Bose–Hubbard result at $t = 10t_{Rabi}$. In contrast to the full many-body result, the Bose–Hubbard wave function remains completely coherent.

a very intricate dynamics results, leading to an equilibration phenomenon, in which the density of the system tends to be equally distributed over both wells.

The system's full many-body dynamics is again strongly fragmented as can be seen in Fig. 6.5(*bottom*), which depicts the natural-orbital occupations $n_i^{(1)}$ for $N = 10$ particles. This rules out any description of the system by Gross–Pitaevskii theory which always predicts condensation. Also shown are the natural-orbital occupations of the Bose–Hubbard model, which wrongly describes a fully condensed system although, in principle, this model can describe fragmentation.

The strong fragmentation of the system implies the presence of strong correlations. This can be seen in the two upper panels of Fig. 6.6, which show the full many-body result for the first-order correlation function $g^{(1)}(x', x; t)$ of $N = 10$ bosons at times $t = 0$ and $t = 10t_{Rabi}$. The fragmentation of the initial state is only $\approx 2\%$, leading to an almost flat $g^{(1)}(x', x; 0)$. This reflects the fact that the system is initially coherent over its entire extent. At $t = 10t_{Rabi}$ the coherence of the system is completely lost even on length scales much shorter than its size, see upper right panel of Fig. 6.6. Note that also $g^{(1)}(x', x; t)$ tends to equilibrate. The respective Bose–Hubbard results for $g^{(1)}(x', x; t)$ are shown in the two lower panels of Fig. 6.6 and in contrast display no visible loss of coherence.

6.8 Conclusions

Let us briefly summarize. We have obtained exact results for the full many-body dynamics of a 1D bosonic Josephson junction. The dynamics is found to be much richer than previously reported. In particular, the predictions of the commonly-employed Gross–Pitaevskii and Bose–Hubbard theories are found to differ substantially from the exact results, already after short times and relatively weak interactions. These differences are associated with the development of fragmentation and correlations not captured by the standard theories. It was found that the more self-trapping is present in the junction, the more the dynamics is coherent. For stronger interactions, where the standard theories predict coherence and self-trapping, we find a completely different dynamics. The system becomes fragmented, spatial coherence is lost and a long-time equilibration of the junction emerges. We hope our results stimulate experiments.

References

1. K. Sakmann, A.I. Streltsov, O.E. Alon, L.S. Cederbaum, Phys. Rev. Lett **103**, 220601 (2009)
2. M.R. Andrews, C.G. Townsend, H.-J. Miesner, D.S. Durfee, D.M. Kurn, W. Ketterle, Science **275**, 637 (1997)
3. T. Schumm, S. Hofferberth, L.M. Andersson, S. Wildermuth, S. Groth, I. Bar-Joseph, J. Schmiedmayer, P. Krüger, Nat. Phys **1**, 57 (2005)
4. C. Orzel, A.K. Tuchman, M.L. Fenselau, M. Yasuda, M.A. Kasevich, Science **291**, 2386 (2001)
5. G.-B. Jo, Y. Shin, S. Will, T.A. Pasquini, M. Saba, W. Ketterle, D.E. Pritchard, Phys. Rev. Lett **98**, 030407 (2007)
6. J. Estève, C. Gross, A. Weller, S. Giovanazzi, M.K. Oberthaler, Nature **455**, 1216 (2008)
7. B.V. Hall, S. Whitlock, R. Anderson, P. Hannaford, A.I. Sidorov, Phys. Rev. Lett **98**, 030402 (2007)
8. G.J. Milburn, J. Corney, E.M. Wright, D.F. Walls, Phys. Rev. A **55**, 4318 (1997)
9. A. Smerzi, S. Fantoni, S. Giovanazzi, S.R. Shenoy, Phys. Rev. Lett **79**, 4950 (1997)
10. M. Albiez, R. Gati, J. Fölling, S. Hunsmann, M. Cristiani, M.K. Oberthaler, Phys. Rev. Lett **95**, 010402 (2005)
11. S. Levy, E. Lahoud, I. Shomroni, J. Steinhauer, Nature (London) **449**, 579 (2007)
12. S. Raghavan, A. Smerzi, S. Fantoni, S.R. Shenoy, Phys. Rev. A **59**, 620 (1999)
13. R. Gati, M.K. Oberthaler, J. Phys. B **40**, R61 (2007)
14. E.A. Ostrovskaya, Y.S. Kivshar, M. Lisak, B. Hall, F. Cattani, D. Anderson, Phys. Rev. A **61**, 031601(R) (2000)

15. Y. Zhou, H. Zhai, R. Lü, Z. Xu, L. Chang, Phys. Rev. A **67**, 043606 (2003)
16. C. Lee, Phys. Rev. Lett **97**, 150402 (2006)
17. D. Ananikian, T. Bergeman, Phys. Rev. A **73**, 013604 (2006)
18. A.N. Salgueiro, de A.F.R. Toledo Piza, G.B. Lemos, R. Drumond, M.C. Nemes, M. Weidemüller, Eur. Phys. J. D **44**, 537 (2007)
19. G. Ferrini, A. Minguzzi, F.W.J. Hekking, Phys. Rev. A **78**, 023606 (2008)
20. V.S. Trippenbach M. Shchesnovich, Phys. Rev. A **78**, 023611 (2008)
21. X.Y. Jia, W.D. Li, J.Q. Liang, Phys. Rev. A **78**, 023613 (2008)
22. M. Trujillo-Martinez, A. Posazhennikova, J. Kroha, Phys. Rev. Lett **103**, 105302 (2009)
23. A.I. Streltsov, O.E. Alon, L.S. Cederbaum, Phys. Rev. Lett **99**, 030402 (2007)
24. O.E. Alon, A.I. Streltsov, L.S. Cederbaum, Phys. Rev. A **77**, 033613 (2008)
25. A.I. Streltsov, O.E. Alon, L.S. Cederbaum, Phys. Rev. A **81**, 022124 (2010)
26. J.C. Eilbeck, P.S. Lomdahl, A.C. Scott, Physica D **16**, 318 (1985)
27. K. Sakmann, A.I. Streltsov, O.E. Alon, L.S. Cederbaum, Phys. Rev. A **78**, 023615 (2008)
28. M. Naraschewski, R.J. Glauber, Phys. Rev. A **59**, 4595 (1999)
29. M. Olshanii, Phys. Rev. Lett **81**, 938 (1998)
30. E.H. Lieb, R. Seiringer, J.P. Solovej, J. Yngvason, *The Mathematics of the Bose Gas and its Condensation* (Birkhäuser Verlag, Basel, 2000)
31. E.H. Lieb, R. Seiringer, Phys. Rev. Lett **91**, 150401 (2003)
32. C.E. Creffield, Phys. Rev. A **75**, 031607(R) (2007)

Chapter 7
Quantum Dynamics of Attractive Versus Repulsive Bosonic Josephson Junctions: Bose–Hubbard and Full-Hamiltonian Results

The quantum dynamics of one-dimensional bosonic Josephson junctions with attractive and repulsive interparticle interactions is studied using the Bose–Hubbard model and by numerically-exact computations of the full many-body Hamiltonian. A symmetry present in the Bose–Hubbard Hamiltonian dictates an equivalence between the evolution in time of attractive and repulsive Josephson junctions with attractive and repulsive interactions of equal magnitude. The full many-body Hamiltonian does not possess this symmetry and consequently the dynamics of the attractive and repulsive junctions are different. The results of this chapter were first published in Ref. [1].

7.1 Introduction

The quantum dynamics of interacting Bose–Einstein condensates is an active and lively research field [2, 3]. Here, one of the basic problems studied is the dynamics of tunneling of interacting Bose–Einstein condensates in double-wells, which in this context are referred to as bosonic Josephson junctions. The dynamics of bosonic Josephson junctions has drawn much attention both theoretically and experimentally, see, e.g., Refs. [4–19] and references therein.

In this chapter we compare the dynamics of one-dimensional bosonic Josephson junctions with attractive and repulsive interparticle interactions. Explicitly, we compare systems where the *magnitude* of attractive and repulsive interactions is the same. We prepare the interacting bosons in one well, and then monitor the evolution of the systems in time. We compute and analyze the respective dynamics both within the two-site Bose–Hubbard model often employed for this problem as well as within the full Hamiltonian of the systems. The main result of this chapter, shown both analytically and numerically, is that within the Bose-Hubbard model the dynamics of the attractive and repulsive junctions is equivalent. In contrast, the dynamics of attractive and repulsive junctions computed from the full many-body Hamiltonian are different from one another. As a complementary result we provide here for the first

K. Sakmann, *Many-Body Schrödinger Dynamics of Bose–Einstein Condensates*, Springer Theses, DOI: 10.1007/978-3-642-22866-7_7,

time in literature numerically-exact quantum dynamics of an attractive Josephson junction, thus matching our recent calculations on repulsive Josephson junctions [5].

7.2 Theory

7.2.1 The Bose–Hubbard Hamiltonian

We begin with the two-site Bose–Hubbard Hamiltonian Eq. 4.14

$$\hat{H}_{BH}(U) = -J\left(\hat{b}_L^\dagger \hat{b}_R + \hat{b}_R^\dagger \hat{b}_L\right) + \frac{U}{2}\left(\hat{b}_L^\dagger \hat{b}_L^\dagger \hat{b}_L \hat{b}_L + \hat{b}_R^\dagger \hat{b}_R^\dagger \hat{b}_R \hat{b}_R\right) \tag{7.1}$$

for a symmetric double well potential $V(-x) = V(x)$. We remind the reader that the Bose–Hubbard Hamiltonian (7.1) can be derived by choosing $W(x - x') = \lambda_0 \delta(x - x')$ in Eq. 2.6, substituting the approximation,

$$\hat{\Psi}(x) = \hat{b}_L \phi_L(x) + \hat{b}_R \phi_R(x) \tag{7.2}$$

for the field operator, Eq. 2.3, into the second quantized representation (2.8) of the full many-body Hamiltonian

$$\hat{\mathcal{H}} = \int dx \hat{\Psi}^\dagger(x) h(x) \hat{\Psi}(x) + \frac{\lambda_0}{2} \int dx \hat{\Psi}^\dagger(x) \hat{\Psi}^\dagger(x) \hat{\Psi}(x) \hat{\Psi}(x), \tag{7.3}$$

neglecting the off-diagonal interaction terms and choosing the energy origin such that $\epsilon = h_{LL} = h_{RR} = 0$. The orbitals $\phi_{L,R}$ are constructed from the ground and the first excited state of *V(x)* and have been defined in Eq. 4.2. For more details on the Bose–Hubbard model we refer the reader to Sect. 4.3.

7.2.2 Attractive–Repulsive Symmetry of the Bose–Hubbard Hamiltonian

There is an interesting symmetry connecting the Bose-Hubbard Hamiltonian (7.1) with repulsive $\hat{H}(U)$ and attractive $\hat{H}(-U)$ interactions of equal magnitude [20, 21, 22, 23]. Explicitly, defining the unitary operator (transformation)

$$\hat{R} = \left\{\hat{b}_L \to \hat{b}_L, \hat{b}_R \to -\hat{b}_R\right\}, \tag{7.4}$$

it is easy to see that [22, 23]

$$\hat{R}\hat{H}(U)\hat{R} = -\hat{H}(-U). \tag{7.5}$$

What is the impact of the symmetry (7.4) and the resulting relation (7.5) on the evolution in time of attractive and repulsive bosonic Josephson junctions? In order to find this out we will restrict the discussion to a system of N bosons initially prepared as mentioned above in, say, the left well, $|N,0\rangle = \frac{1}{\sqrt{N!}}\left(\hat{b}_L^\dagger\right)^N |vac\rangle$. Its evolution in time is simply given by $e^{-i\hat{H}(U)t}\,|N,0\rangle$.

7.2.3 Symmetry of Observables

As in Chap. 6 we define the "survival probability" of finding the bosons in the left well as a function of time by

$$p_L(t;U) = \frac{1}{N}\int_{-\infty}^{0} dx \left\langle N,0 \left| e^{+i\hat{H}(U)t}\left[\hat{\boldsymbol{\Psi}}^\dagger(x)\hat{\boldsymbol{\Psi}}(x)\right] e^{-i\hat{H}(U)t} \right| N,0\right\rangle, \tag{7.6}$$

where the expression $\langle\ldots\rangle$ is the system's density and accordingly the time-averaged "survival probability" by

$$\bar{p}_L(T;U) = \frac{1}{T}\int_0^T dt p_L(t;U). \tag{7.7}$$

just as in Sect. 6.3. Plugging the expansion of the field operator, Eq. 7.2 into the "survival probability" and after some algebra the final result reads:

$$\begin{aligned} p_L(t;U) = &\left(1-\int_{-\infty}^{0} dx\phi_L^2(x)\right) - \left(1-2\int_{-\infty}^{0} dx\phi_L^2(x)\right)\cdot\frac{1}{N} \\ &\times\left\{\left\langle N,0\left|\cos[\hat{H}(U)t]\hat{b}_L^\dagger\hat{b}_L\cos[\hat{H}(U)t] + \sin[\hat{H}(U)t]\hat{b}_L^\dagger\hat{b}_L\sin[\hat{H}(U)t]\right|N,0\right\rangle\right\}. \end{aligned} \tag{7.8}$$

In obtaining the r.h.s. of (7.8) we made use of the facts that the expectation value of $\hat{b}_L^\dagger\hat{b}_L$ (hermitian operator) is real, and that $\int_{-\infty}^{0} dx\phi_L(x)\phi_R(x) = 0$.

Employing the $\hat{R}$ symmetry (7.4) to the $p_L(t;U)$ matrix element (7.8) and making use of relation (7.5), one immediately finds that

$$p_L(t;-U) = p_L(t;U), \tag{7.9}$$

which concludes our first proof. In other words, the "survival probability" of bosons is identical for attractive and repulsive interactions (of equal magnitude) within the Bose–Hubbard model. We emphasize that the result (7.9) holds at all times t. Therefore, also the respective time-averaged "survival probabilities" are the same within the Bose–Hubbard model:

$$\bar{p}_L(T;U) = \bar{p}_L(T;-U). \tag{7.10}$$

Next, we discuss the impact of the symmetry (7.4) on the eigenvalues of the first order RDM within the two-site Bose–Hubbard model. As discussed in Sect. 2.10 the eigenvalues of the first order RDM of a Bose system determine the extent to which the system is condensed or fragmented [24, 25, 26]. For the two-site Bose–Hubbard problem the first order RDM can be written as a two-by-two matrix:

$$\rho^{(1)}(t;U) = \begin{pmatrix} \rho_{LL}(t;U) & \rho_{LR}(t;U) \\ \rho^*_{LR}(t;U) & \rho_{RR}(t;U) \end{pmatrix}, \tag{7.11}$$

where $\rho_{LL}(t;U) = \left\langle N,0 \left| e^{+i\hat{H}(U)t}\hat{b}^\dagger_L \hat{b}_L e^{-i\hat{H}(U)t} \right| N,0 \right\rangle$, and $\rho_{LR}(t;U)$ and $\rho_{RR}(t;U)$ are given analogously. Plugging the symmetry (7.4) into each of the matrix elements of $\rho^{(1)}(t;U)$ and making use of relation (7.5), one can straightforwardly express the first order RDM for attractive interaction as follows:

$$\rho^{(1)}(t;-U) = \begin{pmatrix} \rho_{LL}(t;U) & -\rho^*_{LR}(t;U) \\ -\rho_{LR}(t;U) & \rho_{RR}(t;U) \end{pmatrix}. \tag{7.12}$$

Obviously, the matrices (7.11) and (7.12) have the same characteristic equation, and hence the same eigenvalues. We have thus shown that, within the Bose–Hubbard model, the eigenvalues of the first order RDM do not depend on the sign of interparticle interaction, which constitutes our second proof. Again, this result holds for any time t. In particular, the Bose–Hubbard model attributes identical condensation and fragmentation levels to attractive and repulsive systems.

7.3 Bose–Hubbard Versus Full Hamiltonian Exact Results

To illustrate the above findings we plot in Figs. 7.1 and 7.2 the "survival probability" and occupation numbers, respectively, as a function of time for repulsive and attractive interactions. As a system we choose the same double-well potential $V(x)$ as in Chap. 6, formed by connecting two harmonic potentials $V_\pm(x) = \frac{1}{2}(x \pm 2)^2$ with a cubic spline in the region $|x| \leq 0.5$. The Rabi oscillation period is $t_{Rabi} = \pi/J = 140.66$. As an interaction strength we choose $|\lambda_0| = 0.0129$, leading to $\frac{|U|}{J} = 0.226$. With these parameter values we monitor dynamics of $N = 20$ bosons.

As predicted by Eq. 7.9 and Eqs. 7.11, 7.12, the Bose–Hubbard dynamics for attractive and repulsive junctions are identical. For the time-averaged "survival probability" we find $\bar{p}_L(3t_{Rabi}; \pm U) = 0.74$.

We now move on to the dynamics computed with the full many-body Hamiltonian $\hat{\mathcal{H}}$. In Chap. 6, we presented the first numerically-exact solution of a 1D repulsive bosonic Josephson junction. This exact solution allowed us to unveil novel phenomena associated with the quick loss of the junction's coherence [5]. As in Chap. 6 we use here the MCTDHB method [27, 28, 29] to compute the exact solution

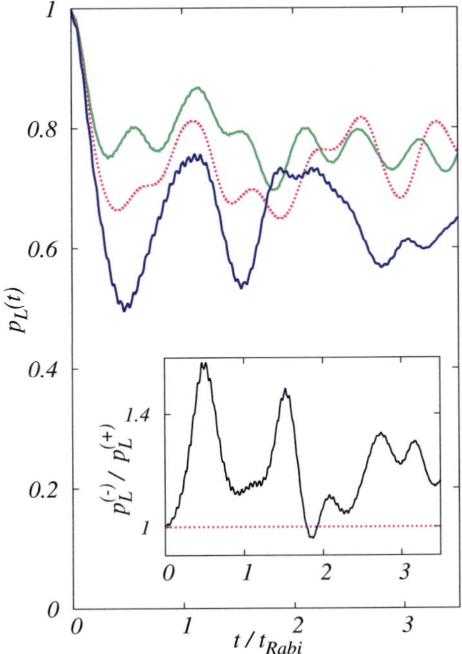

Fig. 7.1 Bose–Hubbard versus full-Hamiltonian, numerically-exact dynamics of attractive and repulsive Josephson junctions. Shown is the "survival probability" as a function of time, $p_L(t)$, computed with the full many-body Hamiltonian for attractive (solid green line) and repulsive (solid blue line) interaction. MCTDHB with $M = 8$ orbitals was used for the full many-body results. The Bose–Hubbard result (dashed magenta line) is for *both* attractive and repulsive interactions. The parameters used are $N = 20$, $\frac{|U|}{J} = 0.226$, $|\lambda_0| = 0.0129$, and $t_{Rabi} = 140.66$. The inset shows the ratio $p_L^{(-)}/p_L^{(+)}$ of attractive to repulsive "survival probabilities" as a function of time. The black–solid line is the full-Hamiltonian results which exhibit a complex dynamics, whereas the dashed–magenta line is the Bose–Hubbard result, showing no dynamics at all. All quantities are dimensionless

of the time-dependent many-body Schrödinger equation using the full many-body Hamiltonian $\hat{\mathcal{H}}$. The MCTDHB method is explained in Sect. 3.1.

Solving the time-dependent many-body Schrödinger equation with the MCTDHB method allows us to report here the first numerically-exact results in literature on a bosonic Josephson junction for attractive interaction, thus matching our recent calculations on repulsive bosonic Josephson junctions [5]. The results of the computations with the full many-body Hamiltonian are collected in Figs. 7.1 and 7.2. It is clearly seen that the dynamics of the attractive and repulsive junctions are distinct from each other, and that each is different from the Bose-Hubbard dynamics. For the time-averaged "survival probability" we find $\bar{p}_L(3t_{Rabi}; U) = 0.66$ for repulsive and $\bar{p}_L(3t_{Rabi}; -U) = 0.79$ for attractive interaction in contrast to the Bose–Hubbard result $\bar{p}_L(3t_{Rabi}; \pm U) = 0.74$.

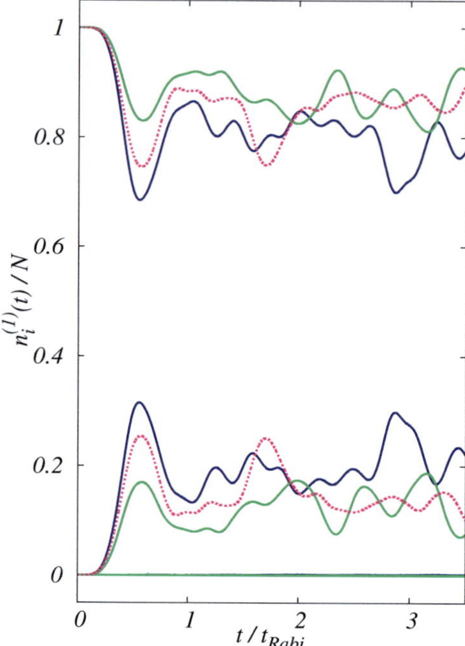

Fig. 7.2 Bose–Hubbard versus full-Hamiltonian, numerically-exact dynamics of attractive and repulsive Josephson junctions. Shown are the occupation numbers of the first order RDM as a function of time, $n_i^{(1)}(t)$, computed with the full many-body Hamiltonian for attractive (solid green line) and repulsive (solid blue line) interaction. The Bose–Hubbard result (dashed magenta line) is for *both* attractive and repulsive interactions. The two-site Bose–Hubbard dynamics has two occupation numbers only. The full-Hamiltonian dynamics has many occupation numbers. It is seen that the occupation numbers $n_{i>2}^{(1)}(t)$ are essentially zero. The parameters used are the same as in Fig. 7.1. All quantities are dimensionless

Let us analyze these findings. We first note, in the context of the above analytical results on the Bose–Hubbard dynamics, that the full Hamiltonian $\hat{\mathcal{H}}$ does not possesses the symmetry (7.4) connecting the dynamics of attractive and repulsive systems. This is because $\hat{\mathcal{H}}$ contains off-diagonal interaction terms as well as all other terms neglected in the Bose–Hubbard Hamiltonian (7.1). From this "mathematical" discussion alone, we do not expect the dynamics of attractive and repulsive junctions to be equivalent as found above for the Bose–Hubbard dynamics. What do we expect on physical grounds? Intuitively, we know that attractive bosons like to be together, whereas repulsive bosons tend to repel each other. These tendencies are exactly what we see in Fig. 7.1. The full-Hamiltonian's "survival probability" is larger (smaller) for attractive (repulsive) interaction than the Bose–Hubbard "survival probability", at least up to $t/t_{Rabi} = 1.5$. In other words, the Bose–Hubbard "survival probability" underestimates the "survival probability" for attractive and overestimates it for repulsive interaction for short and intermediate times. For longer times, as can be seen in Fig. 7.1, the dynamics becomes more complex and anticipating the differences between the exact and the Bose-Hubbard dynamics cannot rest on the above-mentioned physical intuition alone. Finally, Fig. 7.2 presents a complementary picture of the dynamics of occupation numbers. It has been shown in [5] that the Bose–Hubbard dynamics underestimates fragmentation and overestimates coherence of repulsive bosonic Josephson junctions. We may analogously anticipate that the reverse happens with attractive interactions, which indeed is the physical picture unveiled in Fig. 7.2.

Let us conclude. We have shown, both analytically and numerically, that a symmetry present in the two-site Bose–Hubbard Hamiltonian dictates an equivalence between the evolution in time of attractive and repulsive bosonic Josephson junctions. The full many-body Hamiltonian does not possess this symmetry and consequently the dynamics of the attractive and repulsive junctions are distinct. The Bose–Hubbard dynamics underestimates the "survival probability" and overestimates fragmentation of attractive one-dimensional bosonic Josephson junctions and the reverse is true for repulsive junctions. Note that the parameters used here are within the range of expected validity of the Bose–Hubbard model for Josephson junctions [4]. The clear deviations from the numerically-exact results show that criteria for the validity of the Bose–Hubbard model which have been derived for static junctions cannot be transferred for dynamically evolving junctions (see [5]). The present investigation of attractive versus repulsive junctions sheds additional light on the restrictions of the Bose–Hubbard model to describe dynamics. As an outlook, we mention that an analogous symmetry to (7.4) can be found for the multi-site Bose–Hubbard model. Consider the multi-site one-dimensional Bose–Hubbard model:

$$\hat{H}_{BH}(U) = -J\sum_j (\hat{b}_j^\dagger \hat{b}_{j+1} + \hat{b}_{j+1}^\dagger \hat{b}_j) + \frac{U}{2}\sum_j \hat{b}_j^\dagger \hat{b}_j^\dagger \hat{b}_j \hat{b}_j. \tag{7.13}$$

Then,

$$\hat{R}_{BH}\hat{H}_{BH}(U)\hat{R}_{BH} = -\hat{H}_{BH}(-U) \tag{7.14}$$

where $\hat{R}_B H = \left\{\hat{b}_{2j} \rightarrow \hat{b}_{2j}, \hat{b}_{2j-1} \rightarrow -\hat{b}_{2j-1}\right\}$. The extension to the Bose–Hubbard model of orthorhombic lattices in higher dimensions is straightforward. It would be interesting to search for the consequences of this symmetry in the dynamics of attractive and repulsive bosons in a lattice.

References

1. K. Sakmann, A.I Streltsov, O.E. Alon, L.S. Cederbaum, Phys. Rev. A **82**, 013620 (2010)
2. L. Pitaevskii, S. Stringari, Bose-Einstein Condensation. (Oxford University Press, Oxford, 2003)
3. C.J. Pethick, H. Smith, Bose-Einstein Condensation in Dilute Gases. 2nd edn. (Cambridge University Press, Cambridge, 2008)

4. G.J. Milburn, J. Corney, E.M. Wright, D.F. Walls, Phys. Rev. A **55**, 4318 (1997)
5. K. Sakmann, A.I. Streltsov, O.E. Alon, L.S. Cederbaum, Phys. Rev. Lett **103**, 220601 (2009)
6. A. Smerzi, S. Fantoni, S. Giovanazzi, S.R. Shenoy, Phys. Rev. Lett **79**, 4950 (1997)
7. S. Raghavan, A. Smerzi, S. Fantoni, S.R. Shenoy, Phys. Rev. A **59**, 620 (1999)
8. R. Gati, M.K. Oberthaler, J. Phys. B **40**, R61 (2007)
9. M. Albiez, R. Gati, J. Fölling, S. Hunsmann, M. Cristiani, M.K. Oberthaler, Phys. Rev. Lett **95**, 010402 (2005)
10. S. Levy, E. Lahoud, I. Shomroni, J. Steinhauer, Nature (London), **449**, 579 (2007)
11. E.A. Ostrovskaya, Y.S. Kivshar, M. Lisak, B. Hall, F. Cattani, D. Anderson, Phys. Rev. A **61**, 031601(R) (2000)
12. Y. Zhou, H. Zhai, R. Lü, Z. Xu, L. Chang, Phys. Rev. A **67**, 043606 (2003)
13. C. Lee, Phys. Rev. Lett **97**, 150402 (2006)
14. D. Ananikian, T. Bergeman, Phys. Rev. A **73**, 013604 (2006)
15. A.N. Salgueiro, de A.F.R. Toledo Piza, G.B. Lemos, R. Drumond, M.C. Nemes, M. Weidemüller, Eur. Phys. J. D **44**, 537 (2007)
16. G. Ferrini, A. Minguzzi, F.W.J. Hekking, Phys. Rev. A **78**, 023606 (2008)
17. V.S. Shchesnovich, M. Trippenbach, Phys. Rev. A **78**, 023611 (2008)
18. X.Y. Jia, W.D. Li, J.Q. Liang, Phys. Rev. A **78**, 023613 (2008)
19. M. Trujillo-Martinez, A. Posazhennikova, J. Kroha, Phys. Rev. Lett **103**, 105302 (2009)
20. R. Franzosi, V. Penna, Phys. Rev. A **63**, 043609 (2001)
21. P. Buonsante, R. Franco, V. Penna, J. Phys. A **38**, 8393 (2005)
22. J. Links, K.E. Hibberd, SIGMA **2** (2006). Paper 095
23. J. Links, S.-Y. Zhao, J. Stat. Mech (2009) P03013
24. A.I. Streltsov, O.E. Alon, L.S. Cederbaum, Phys. Rev. A **73**, 063626 (2006)
25. E.J. Mueller, T.-L. Ho, M. Ueda, G. Baym, Phys. Rev. A **74**, 033612 (2006)
26. K. Sakmann, A.I. Streltsov, O.E. Alon, L.S. Cederbaum, Phys. Rev. A **78**, 023615 (2008)
27. A.I. Streltsov, O.E. Alon, L.S. Cederbaum, Phys. Rev. Lett **99**, 030402 (2007)
28. O.E. Alon, A.I. Streltsov, L.S. Cederbaum, Phys. Rev. A **77**, 033613 (2008)
29. O.E. Alon, A.I. Streltsov, L.S. Cederbaum, J. Chem. Phys **127**, 154103 (2007)

Chapter 8
Optimal Time-Dependent Lattice Models for Nonequilibrium Dynamics

A method to extend and optimize lattice models is presented. While conventional lattice models, such as the Bose–Hubbard (BH) model, use Wannier functions that are time-independent, we allow the Wannier functions of lattice models to depend on time. By deriving equations of motion for these new lattice models from the variational principle we show that it is possible to exploit the additional freedom introduced by allowing the Wannier functions to depend on time, to improve the lattice model. Since all parameters are then determined by the variational principle the lattice model becomes optimal. As an example we show this explicitly for the BH model in a quantum quench scenario. The improvement due to the use of time-dependent Wannier functions is quantified by comparison with exact results of the time-dependent many-body Schrödinger equation. The concept is general and can be applied to any lattice model that employs Wannier functions. The results of this Chapter were first published in Ref. [1].

8.1 Introduction

At the heart of many lattice models is the idea of lattice site localized functions which are commonly referred to as Wannier functions. Strictly speaking Wannier functions are only defined for purely periodic potentials, but the term is used more generally. Wannier functions are constructed from (energetically) nearly-degenerate eigenfunctions of a one-body problem, e.g., the Bloch waves in strictly periodic potentials or the doublets in double-wells, see Chap. 4. In lattice models the Hamiltonian in its second-quantized form is expanded in the basis Wannier functions used. The fact that even the Wannier functions of adjacent lattice sites have little overlap motivates the neglect of terms in the Hamiltonian. Lattice models are therefore classified according to how many terms are kept. For example, the number of bands used distinguishes single-band from multi-band models and the coupling to neighboring lattice sites determines whether the model contains nearest-neighbor interactions, next nearest-neighbor interactions and so on. Most models are restricted

K. Sakmann, *Many-Body Schrödinger Dynamics of Bose–Einstein Condensates*,
Springer Theses, DOI: 10.1007/978-3-642-22866-7_8,

to a single band and no more than next nearest-neighbor interactions, which is of great advantage in numerical and analytical computations. Eventually an appealing picture of particles hopping from one lattice site to another emerges. Probably the most prominent of such lattice models are the Hubbard model [2–7] for fermions and its bosonic counterpart, the BH model [8], both of which are single-band models with nearest-neighbor interaction.

The restriction to a single band and the neglect of terms in the Hamiltonian simplifies the solution of lattice models, but makes them also less accurate. In fact, it has recently become clear that single-band models can be far too restrictive and can miss the true physics [9–13] even when the parameters are chosen such that certain validity criteria are fulfilled, see Refs. [9, 10] and Chap. 6. Due to a lack of reliable validity criteria it is unfortunately not clear what the range of validity of such models really is. One way to answer this question is to compare their results to exact solutions of the many-body Schrödinger equation, as done in Chap. 6 using the MCTDHB method. Another option would be to successively include more of the previously neglected terms and bands until convergence is reached. This second path is restricted to systems of small size if conventional Wannier functions are used, because the number of basis functions grows quickly beyond what is computationally feasible. A straightforward solution to this dilemma is to include only certain classes of many-body basis functions in a computation, e.g., by enforcing an energy cutoff on the many-body basis. Unfortunately, such procedures lack size-consistency [14, 15], a crucial requirement if the obtained results are to be extrapolated to larger systems, which is usually desired. It is therefore important to examine the possibilities to make the best possible choice for the Wannier functions, in order to minimize the number of bands needed for convergence.

In the following sections we show that it is possible to greatly improve lattice models by letting the Wannier functions become time-dependent. Using the principle of least action we will derive equations of motion for these time-dependent Wannier functions that ensure the optimal dynamics within the given model. Of course also stationary states that are variationally optimal within the model can be obtained, and thereby the phase diagram. The concept of *time-dependent Wannier functions* presented here is general and can be applied to fermions and bosons alike and any number of lattice sites, particles and bands. However, for simplicity we will illustrate it for the BH model only. When the Wannier functions of the BH model are allowed to depend on time, we will refer to it as the *time-dependent Bose–Hubbard* (TDBH) model, even though the underlying many-body Hamiltonian itself does not depend on time. The thereby obtained gain in accuracy will be shown explicitly by considering a quantum quench scenario.

8.2 Theory

The number of many-body basis functions in a lattice model is a rapidly growing function of the number of particles N, the number of lattice sites M and the number of bands κ. For example, in a bosonic model the number of basis functions is then

$\binom{N+\kappa M-1}{N}$ which grows factorially with any of the parameters. This demonstrates the necessity to optimize the Wannier functions used. We have included an example that demonstrates this necessity in Appendix F. For simplicity, we restrict the discussion to the case of a 1D lattice consisting of $M=2$ sites, N bosons and a single band, $\kappa=1$. The number of basis functions is then just $\binom{N+1}{N}=N+1$.

Our starting point is the following *ansatz* for the two *time-dependent Wannier functions*

$$\phi_k(x,t)=\sum_{\alpha=1}^{\nu} d_{k_\alpha}(t)\phi_{k_\alpha}(x), k=L,R \tag{8.1}$$

where $d_{k_\alpha}(t)$ are *local orbital coefficients* and the functions $\{\phi_{L_\alpha}\}$, $\alpha=1,\ldots,\nu$ and $\{\phi_{R_\beta}\}$, $\beta=1,\ldots,\nu$ form two mutually orthogonal sets of functions

$$\langle\phi_{k_\alpha}|\phi_{j_\beta}\rangle=\delta_{kj}\delta_{\alpha\beta}, k=L,R, \tag{8.2}$$

e.g., the time-independent Wannier functions of the bands $\alpha=1,\ldots,\nu$. The ansatz (8.1) then automatically satisfies

$$\langle\phi_L|\phi_R\rangle=0 \tag{8.3}$$

at all times. Normalization requires that the local orbital coefficients $d_{k\alpha}(t)$ satisfy

$$\sum_{\alpha=1}^{\nu}|d_{k_\alpha}(t)|^2=\langle\phi_k|\phi_k\rangle=1, \quad k=L,R \tag{8.4}$$

The ansatz for the two-mode many-body wave function thereby becomes

$$\begin{aligned}|\Psi(t)\rangle&=\sum_{\vec{n}}C_{\vec{n}}(t)\,|\vec{n};t\rangle,\\ |\vec{n};t\rangle=|n_L,n_R;t\rangle&=\frac{1}{\sqrt{n_L!n_R!}}\left(\hat{b}_L^\dagger(t)\right)^{n_L}\left(\hat{b}_R^\dagger(t)\right)^{n_R}|vac\rangle,\end{aligned} \tag{8.5}$$

where $n_L+n_R=N$ and the sum runs over all $N+1$ time-dependent permanents $|n_L,n_R;t\rangle$. With this *ansatz* the finite size representation, Eq. 2.16, of the field operator, Eq. 2.3, takes on the form

$$\hat{\Psi}_M(x;t)=\hat{b}_L(t)\phi_L(x,t)+\hat{b}_R(t)\phi_R(x,t). \tag{8.6}$$

By substituting (2)into the expression for the full many-body Hamiltonian, Eq. 8.6, we arrive at the representation

$$\hat{H}_{TD2M}=\sum_{k,q=L,R}\hat{b}_k^\dagger(t)\hat{b}_q(t)h_{kq}(t)+\frac{1}{2}\sum_{k,s,l,q=L,R}\hat{b}_k^\dagger(t)\hat{b}_s^\dagger(t)\hat{b}_l(t)\hat{b}_q(t)W_{ksql}(t), \tag{8.7}$$

of the full two-mode Hamiltonian, with the matrix elements defined in Eq. 2.10. It will be useful to note that the interaction matrix elements can always be written as

$$W_{ksql}(t) = \langle \phi_k | \hat{W}_{sl} | \phi_q \rangle \tag{8.8}$$

using the definition (3.7) of $W_{sl}(x,t)$. As an interaction potential we choose $W(x-x') = \lambda_0 \delta(x-x')$ and define

$$\begin{aligned} J(t) &= -h_{LR}(t) = -h_{RL}(t)^*, \\ \epsilon_k(t) &= h_{kk}(t), \\ U_{kk}(x,t) &= W_{kk}(x,t) \\ U_{kkkk}(t) &= \langle \phi_k | \hat{U}_{kk} | \phi_k \rangle, \end{aligned} \tag{8.9}$$

for $k = L, R$. We note that if only the Wannier functions of the lowest band are used, $\nu = 1$ in the ansatz (8.1) for the time-dependent Wannier functions, all of the time-dependent parameters defined above reduce to those of the time-independent BH model, $U_{LLLL}(t) = U_{RRRR} = U$, $J(t) = J$, $\epsilon_L(t) = \epsilon_R(t) = \epsilon$, as defined in Sect. 4.1. Neglecting all off-diagonal terms in the two-body part of Eq. 8.7 we arrive at the TDBH Hamiltonian

$$\begin{aligned} \hat{H}_{TDBH} = & -J(t)\hat{b}_L^\dagger(t)\hat{b}_R(t) - J^*(t)\hat{b}_R^\dagger(t)\hat{b}_L(t) \\ & + \epsilon_L(t)\hat{b}_L^\dagger(t)\hat{b}_L(t) + \epsilon_R(t)\hat{b}_R^\dagger(t)\hat{b}_R(t) \\ & + \frac{1}{2}U_{LLLL}(t)\hat{b}_L^\dagger(t)\hat{b}_L^\dagger(t)\hat{b}_L(t)\hat{b}_L(t) \\ & + \frac{1}{2}U_{RRRR}(t)\hat{b}_R^\dagger(t)\hat{b}_R^\dagger(t)\hat{b}_R(t)\hat{b}_R(t). \end{aligned} \tag{8.10}$$

We will now derive equations of motion for the coefficients $\{C_{\vec{n}}(t)\}$ and the time-dependent Wannier functions $\phi_{L,R}(x,t)$ by requiring stationarity of the action integral

$$S\left[\{C_{\vec{n}}\}, \{d_{k_\alpha}\}\right] = \int dt \left\{ \left\langle \Psi \left| \hat{H}_{TDBH} - i\frac{\partial}{\partial t} \right| \Psi \right\rangle - \sum_{m,n=L,R} \mu_{mn}(t) \left[\left\langle \phi_m | \phi_n \right\rangle - \delta_{mn} \right] \right\} \tag{8.11}$$

with respect to variations of the coefficients $\{C_{\vec{n}}(t)\}$ and the local orbital coefficients $\{d_{k\alpha}(t)\}$. Here we have used the Hamiltonian $\hat{H}_{TDBH}$ in *S*, but of course also the full two-mode Hamiltonian $\hat{H}_{TD2M}$ can be used. Note that the Lagrange multipliers μ_{LR}, μ_{RL} are actually superfluous because Eq. 8.3 already ensures orthogonality between the orbitals ϕ_L and ϕ_R by construction.

8.2.1 Variation with Respect to the Coefficients $C^*_{\vec{n}}(t)$

To take the variation with respect to the expansion coefficients $C^*_{\vec{n}}(t)$ we rewrite the first part of the integrand of (8.11) as

$$\left\langle \Psi \left| \hat{H}_{TDBH} - i\frac{\partial}{\partial t} \right| \Psi \right\rangle = \sum_{\vec{n}} C^*_{\vec{n}} \left[\sum_{\vec{n}'} \left\langle \vec{n}; t \left| \hat{H}_{TDBH} - i\frac{\partial}{\partial t} \right| \vec{n}'; t \right\rangle C_{\vec{n}'} - i\frac{\partial C_{\vec{n}}}{\partial t} \right] \quad (8.12)$$

and require stationarity of the action integral (2) with respect to the coefficients $C^*_{\vec{n}}(t)$:

$$0 = \frac{\delta S}{\delta C^*_{\vec{n}}(t)} \Longleftrightarrow \sum_{\vec{n}'} \left\langle \vec{n}; t \left| \hat{H}_{TDBH} - i\frac{\partial}{\partial t} \right| \vec{n}'; t \right\rangle C_{\vec{n}'} = i\frac{\partial C_{\vec{n}}}{\partial t} \quad (8.13)$$

Defining the time-dependent matrix $\mathcal{H}(t)$ by

$$\mathcal{H}_{\vec{n}\vec{n}'}(t) = \left\langle \vec{n}; t \left| \hat{H}_{TDBH} - i\frac{\partial}{\partial t} \right| \vec{n}'; t \right\rangle. \quad (8.14)$$

and collecting all coefficients $C_{\vec{n}}(t)$ in a vector $\mathbf{C}(t)$, Eq. 8.12 takes on the form

$$\mathcal{H}(t)\mathbf{C}(t) = i\frac{\partial \mathbf{C}(t)}{\partial t}. \quad (8.15)$$

The time-evolution of the coefficient vector $\mathbf{C}$ is unitary if the matrix $\mathcal{H}(t)$ is Hermitian. Since the Hamiltonian $\hat{H}_{TDBH}$ is a hermitian operator the only part that needs to be discussed is $\left\langle \vec{n}; t \left| i\frac{\partial}{\partial t} \right| \vec{n}'; t \right\rangle$. When acting on permanents the time derivative can be written as

$$i\frac{\partial}{\partial t} = \sum_{k,q=L,R} \hat{b}^\dagger_k \hat{b}_q \left(i\frac{\partial}{\partial t} \right)_{kq}, \quad (8.16)$$

where

$$\left(i\frac{\partial}{\partial t} \right)_{kq} = i \int dx \phi_k(x,t)^* \frac{\partial \phi_q(x,t)}{\partial t} \quad (8.17)$$

The normalization condition $i\frac{d}{dt}\langle \phi_k | \phi_q \rangle = 0$, then implies that $\left(i\frac{\partial}{\partial t}\right)_{kq} = \left(i\frac{\partial}{\partial t}\right)^*_{qk}$ and hence that the matrix $\mathcal{H}$ is hermitian. The time-evolution of the coefficient vector $\mathbf{C}$ is therefore unitary. Note that Eq. 8.2 implies

$$\left(i\frac{\partial}{\partial t} \right)_{LR} = \left(i\frac{\partial}{\partial t} \right)_{RL} = 0 \quad (8.18)$$

and hence the expansion (8.16) can only have non-zero matrix elements for $k = q$, i.e., on the diagonal.

8.2.2 Variation with Respect to the Local Orbital Coefficients $d^*_{k_\alpha}(t)$

In order to derive equations of motion for the time-dependent Wannier functions it is helpful to express the expectation value $\left\langle \Psi \left| \hat{H}_{TDBH} - i\frac{\partial}{\partial t} \right| \Psi \right\rangle$ in a form that allows for a direct functional derivative with respect to $d^*_{k_\alpha}(t)$. Defining

$$
\begin{aligned}
h_{k_\alpha q_\beta} &= \left\langle \phi_{k_\alpha} \left| \hat{h} \right| \phi_{q_\beta} \right\rangle \\
U_{k_\alpha k_\beta k_\gamma k_\delta} &= \lambda_0 \int dx \phi^*_{k_\alpha}(x) \phi^*_{k_\beta}(x) \phi_{k_\delta}(x) \phi_{k_\delta}(x)
\end{aligned}
\tag{8.19}
$$

and using the expressions (2.33) and (2.34) for the first and second order RDMs we find

$$
\begin{aligned}
\left\langle \Psi \left| \hat{H}_{TDBH} - i\frac{\partial}{\partial t} \right| \Psi \right\rangle = & \sum_{k,q} \sum_{\alpha,\beta}^{\nu} \rho_{kq}(t) d^*_{k_\alpha}(t) \left(h_{k_\alpha q_\beta} - i \delta_{kq} \delta_{\alpha\beta} \frac{\partial}{\partial t} \right) d_{q_\beta}(t) \\
& + \frac{1}{2} \sum_{k} \sum_{\alpha\beta\gamma\delta} \rho_{kkkk}(t) d^*_{k_\alpha}(t) d^*_{k_\beta}(t) U_{k_\alpha k_\beta k_\gamma k_\delta} d_{k_\gamma}(t) d_{k_\delta}(t),
\end{aligned}
\tag{8.20}
$$

We write $\frac{\partial}{\partial t}\phi_q(x,t) \equiv \dot{\phi}_q(x,t)$ and note that $d_{k_\alpha}(t) = \langle \phi_{k_\alpha} | \phi_q \rangle$. Requiring stationarity of the action with respect to variations of the local orbital coefficients, $0 = \frac{\delta S}{\delta d^*_{k_\alpha}(t)}$, results in

$$
\begin{aligned}
\sum_{q=L,R} \rho_{Lq} \langle \phi_{L_\alpha} | \hat{h} | \phi_q \rangle + \rho_{LLLL} \langle \phi_{L_\alpha} | \hat{U}_{LL} | \phi_L \rangle &= \mu_{LL} \langle \phi_{L_\alpha} | \phi_L \rangle + i \rho_{LL} \langle \phi_{L_\alpha} | \dot{\phi}_L \rangle \\
\sum_{q=L,R} \rho_{Rq} \langle \phi_{R_\alpha} | \hat{h} | \phi_q \rangle + \rho_{RRRR} \langle \phi_{R_\alpha} | \hat{U}_{RR} | \phi_R \rangle &= \mu_{RR} \langle \phi_{R_\alpha} | \phi_R \rangle + i \rho_{RR} \langle \phi_{R_\alpha} | \dot{\phi}_R \rangle
\end{aligned}
\tag{8.21}
$$

for $\alpha = 1, \ldots, \nu$. As expected on the basis of Eq. 8.3, the superfluous Lagrange multipliers μ_{LR} and μ_{RL} do not appear in the set of equations (8.21). The remaining Lagrange multipliers μ_{LL} and μ_{RR} can be determined from Eq. 8.21 as

$$
\begin{aligned}
\mu_{LL} &= \sum_{\alpha=1}^{\nu} \langle \phi_L | \phi_{L_\alpha} \rangle \mu_{LL} \langle \phi_{L_\alpha} | \phi_L \rangle \\
&= \rho_{LL} \left(h_{LL} - i \langle \phi_L | \dot{\phi}_L \rangle \right) + \rho_{LR} h_{LR} + \rho_{LLLL} U_{LLLL}
\end{aligned}
\tag{8.22}
$$

and analogously

$$
\begin{aligned}
\mu_{RR} &= \sum_{\alpha=1}^{\nu} \langle \phi_R | \phi_{R_\alpha} \rangle \mu_{RR} \langle \phi_{R_\alpha} | \phi_R \rangle \\
&= \rho_{RR} \left(h_{RR} - i \langle \phi_R | \dot{\phi}_R \rangle \right) + \rho_{RL} h_{RL} + \rho_{RRRR} U_{RRRR}
\end{aligned}
\tag{8.23}
$$

We now define the projectors

$$\hat{P}_k = \sum_{\beta=1}^{\nu} |\phi_{k_\beta}\rangle\langle\phi_{k_\beta}| - |\phi_k\rangle\langle\phi_k|, \quad k = L, R \tag{8.24}$$

and substitute the expressions (8.22), (8.23) for the Lagrange multipliers into (8.21) and obtain

$$\begin{aligned} i\rho_{LL}\langle\phi_{L_\alpha}|\hat{P}_L|\dot{\phi}_L\rangle &= \langle\phi_{L_\alpha}|\hat{P}_L\left[\sum_{q=L,R}\rho_{Lq}\hat{h}|\phi_q\rangle + \rho_{LLLL}\hat{U}_{LL}|\phi_L\rangle\right] \\ i\rho_{RR}\langle\phi_{R_\alpha}|\hat{P}_R|\dot{\phi}_R\rangle &= \langle\phi_{R_\alpha}|\hat{P}_R\left[\sum_{q=L,R}\rho_{Rq}\hat{h}|\phi_q\rangle + \rho_{RRRR}\hat{U}_{RR}|\phi_R\rangle\right]. \end{aligned} \tag{8.25}$$

Multiplying each of these two equations from the left with $\sum_\alpha |\phi_{k_\alpha}\rangle$ results in

$$\begin{aligned} i\rho_{LL}\hat{P}_L|\dot{\phi}_L\rangle &= \hat{P}_L\left[\sum_{q=L,R}\rho_{Lq}\hat{h}|\phi_q\rangle + \rho_{LLLL}\hat{U}_{LL}|\phi_L\rangle\right] \\ i\rho_{RR}\hat{P}_R|\dot{\phi}_R\rangle &= \hat{P}_R\left[\sum_{q=L,R}\rho_{Rq}\hat{h}|\phi_q\rangle + \rho_{RRRR}\hat{U}_{RR}|\phi_R\rangle\right]. \end{aligned} \tag{8.26}$$

In this form the equations for the time-dependent Wannier functions of the BH model are hard to solve. Fortunately, it is possible to simplify Eq. 8.26 by shifting the phase of ϕ_L and ϕ_R as follows. We define

$$|\tilde{\phi}_k\rangle \equiv e^{-\int^t dt'\langle\phi_k|\dot{\phi}_k\rangle}|\phi_k\rangle \tag{8.27}$$

for $k = L, R$. In order to indicate that $\tilde{\phi}_k$ is to be used instead of ϕ_k in any matrix element or operator we will write $\tilde{k}$ instead of k in the following. Using the fact that $\langle\phi_k|\dot{\phi}_k\rangle = -\langle\phi_k|\dot{\phi}_k\rangle^*$ it is then easy to verify that

$$\begin{aligned} \hat{P}_k|\dot{\phi}_k\rangle &= e^{+\int^t dt'\langle\phi_k|\dot{\phi}_k\rangle}|\dot{\tilde{\phi}}_k\rangle \\ \rho_{kq}\hat{h}|\phi_q\rangle &= e^{+\int^t dt'\langle\phi_k|\dot{\phi}_k\rangle}\rho_{\tilde{k}\tilde{q}}\hat{h}|\phi_{\tilde{q}}\rangle \\ \rho_{kkkk}\hat{U}_{kk}|\phi_k\rangle &= e^{+\int^t dt'\langle\phi_k|\dot{\phi}_k\rangle}\rho_{\tilde{k}\tilde{k}\tilde{k}\tilde{k}}\hat{U}_{\tilde{k}\tilde{k}}|\phi_{\tilde{k}}\rangle \\ \hat{P}_k &= \hat{P}_{\tilde{k}}. \end{aligned} \tag{8.28}$$

Substituting Eqs. 8.28 into 8.26 and dropping the tilde we thereby arrive at a more tractable form of the orbital equations of the TDBH model

$$\begin{aligned} i|\dot{\phi}_L\rangle &= \hat{P}_L\left[\hat{h}|\phi_L\rangle + \frac{\rho_{LR}}{\rho_{LL}}\hat{h}|\phi_R\rangle + \frac{\rho_{LLLL}}{\rho_{LL}}\hat{U}_{LL}|\phi_L\rangle\right] \\ i|\dot{\phi}_R\rangle &= \hat{P}_R\left[\hat{h}|\phi_R\rangle + \frac{\rho_{RL}}{\rho_{RR}}\hat{h}|\phi_L\rangle + \frac{\rho_{RRRR}}{\rho_{RR}}\hat{U}_{RR}|\phi_R\rangle\right]. \end{aligned} \tag{8.29}$$

The equations of motion (8.29) ensure that at each time step the change of any orbital is orthogonal to itself. The solution of the equations (8.29) is much simpler than that of Eq. 8.26 because projectors appear now only on the right-hand side. Furthermore, Eq. 8.29 imply that

$$\left(i\frac{\partial}{\partial t}\right)_{LL} = \left(i\frac{\partial}{\partial t}\right)_{RR} = \left(i\frac{\partial}{\partial t}\right)_{LR} = \left(i\frac{\partial}{\partial t}\right)_{RL} = 0, \tag{8.30}$$

which greatly simplifies the evaluation of the matrix $\mathcal{H}$.

8.2.3 Remarks on the Time-Dependent Bose–Hubbard Model

In the previous section we employed an *ansatz* (8.5) for the many-body wave function, which allowed the Wannier functions to depend on time. The *ansatz* (8.5) is the most general ansatz possible that can be constructed from two lattice site localized orbitals. By employing the principle of least action we then derived coupled equations of motion for the coefficients $\{C_{\vec{n}}\}(t)$ of the many-body wave function and the time-dependent Wannier functions $\phi_L(x,t)$ and $\phi_R(x,t)$. For any given initial state, consisting of coefficients $\{C_{\vec{n}}(0)\}$ and orbitals $\phi_L(x,0)$, $\phi_R(x,0)$ the solution of Eq. 8.15 together with Eq. 8.29 is the variationally optimal answer to the many-boson problem within the framework of the BH Hamiltonian. For the TDBH model the number of local orbital coefficients ν should always be chosen so large that the results do not depend on its precise value. However, if the Wannier functions ϕ_L and ϕ_R in the TDBH model are restricted to the lowest band by setting $\nu = 1$ in Eq. 8.1, the TDBH model reduces to the BH model. As is well known, the BH model has only one relevant parameter, the ratio U/J which is constant in time and within the BH model the dynamics of all N-boson systems with the same ratio U/J is identical. Meanwhile, the TDBH model has four real parameters $U_{LLLL}(t)$, $U_{RRRR}(t)$, $\epsilon_L(t)$ and $\epsilon_R(t)$ and one complex parameter $J(t)$, all of which are time-dependent and are implicitly determined by the variational principle. Thus, time-dependent Wannier functions allow for a much less restricted dynamics within the framework of the BH model, whilst keeping the appealing picture of bosons hopping from one site to another. Another point worth mentioning is the possibility to use time-dependent Wannier functions as a test for the validity of any given lattice model that employs time-independent Wannier functions. As mentioned earlier, the number of many-body basis functions of a bosonic lattice model is $\binom{N+\kappa M-1}{N}$. This number does not grow if each of the κM Wannier functions is allowed to depend on time. The additional cost of letting Wannier functions become time-dependent consists of solving κM equations of motion for the Wannier functions, which depend on the coefficients $\{C_{\vec{n}}(t)\}$ and vice versa. This additional cost is usually small compared to the cost of including another band, which means $\kappa \rightarrow \kappa + 1$. The validity of any given lattice model can therefore be tested by comparing the results using time-dependent and time-independent Wannier functions. If the results differ, the time-independent lattice model is inapplicable. The reverse is of course not necessarily true.

8.2.4 *Implementation of the Time-Dependent Bose–Hubbard Model*

The solution of the TDBH model requires the simultaneous solution of Eqs. 8.29 and (8.15) using Eq. 8.30. The coefficients $\{C_{\vec{n}}\}(t)$ in Eq. 8.15 depend on the time-dependent Wannier functions through the matrix elements $J(t), \epsilon_L(t), \epsilon_R(t)$, $U_{LLLL}(t)$ and $U_{RRRR}(t)$. In turn the equations (8.29) depend on the coefficients $\{C_{\vec{n}}\}(t)$ via the matrix elements ρ_{jk} and ρ_{kkkk} as can be seen from the identities collected in Appendix D. An implementation of this coupled system of equations is possible and proceeds along the same lines as an implementation of the MCTDHB equations [16, 17]. We will not go further into the details of the implementation and refer the reader to Refs. [16, 17].

Obviously, the propagation of Eq. 8.29 is ill-defined as soon as either ρ_{LL} or ρ_{RR} is zero. The density matrix will then have to be renormalized to avoid a singularity. As explained in Ref. [18] such a regularization affects only the propagation of orbitals in which no particles reside and is therefore without effect on observable quantities. This has also been confirmed in numerical tests of the present model. For the double-well system considered here the regularization is necessary if the system is in either of the states $|N, 0\rangle$ or $|0, N\rangle$, i.e., when all particles are in the left or in the right well.

8.3 Example of Dynamics Using Time-Dependent Wannier Functions

In this section we illustrate explicitly how the usage of time-dependent Wannier functions improves a lattice model. We choose a quantum quench scenario in which the interaction strength is suddenly increased from zero to a finite value.

8.3.1 *A Double-Well Potential as a Test System*

In order to quantify the improvement that results from the use of time-dependent Wannier functions as opposed to conventional, time-independent ones, we now turn to a specific example. As an external trapping potential we choose a double-well with periodic boundary conditions

$$V(x) = V_{0x}\cos(kx)^2 \tag{8.31}$$

on the interval $[-\pi, \pi)$ with $k = 1$. The recoil energy for a boson of mass m is then given by

$$E_r = \frac{k^2}{2m}. \tag{8.32}$$

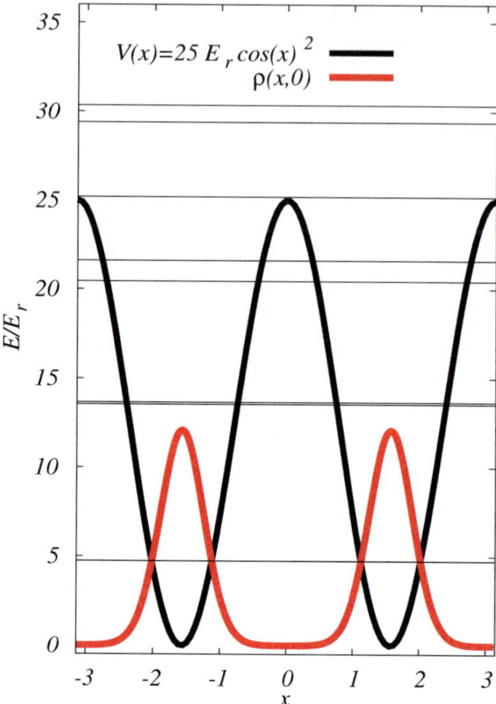

Fig. 8.1 Initial state of the quench dynamics. Shown is the density (*thick red line*) of the initial state in the double well potential $V(x) = 25E_r \cos^2(x)$ (*thick black line*). The initial state is the noninteracting ground state of $N = 20$ bosons. Also shown are the single-particle energy levels (*horizontal black lines*) of the potential. The lowest six energy levels are below the barrier height $V(0) = 25E_r$

We are working in dimensionless units in which $\hbar = m = 1$ and therefore $E_r = 1/2$. As a depth we use $V_{0x} = 25E_r$. In Fig. 8.1 the potential $V(x)$ is shown together with its single-particle energy levels and the density of the noninteracting ground state. The six lowest single-particle energy levels are below the barrier: $e_1 = 4.733E_r$, $e_2 = 4.737E_r$, $e_3 = 13.531E_r$, $e_4 = 13.650E_r$, $e_5 = 20.443E_r$, $e_6 = 21.601E_r$. For the Rabi oscillation period we then find $t_{Rabi} = \pi/J = 3025$. Using, e.g., the parameters of the experiment in Ref. [19], with ^{87}Rb as a boson and a lattice spacing of 600 nm the Rabi oscillation period becomes $t_{Rabi} = 150.9$ ms.

In the following we will study the dynamics of bosons in this potential and compare the results of the TDBH model with those of the BH model and the exact many-body Schrödinger equation, computed using MCTDHB. For the TDBH model we use $\nu = 10$ local orbitals per site, although less (about five) are necessary to obtain convergence within the model.

8.3.2 Quantum Quench Dynamics in a Double-Well Potential

When a quantum system is initially prepared in the ground state of some Hamiltonian the dynamics due to a sudden change in one of the parameters of the Hamiltonian is known as a *quantum quench*. Recently, quantum quenches using ultracold bosons have received a lot of attention in the context of superfluid to Mott insulator transitions [20, 22] as well as thermalization and integrability [23–27].

Here, we consider a quench scenario in the double-well potential $V(x)$ and note that such a quench can be implemented experimentally by using Feshbach resonances. As an initial state we use $N = 20$ bosons in the noninteracting ground state of $V(x)$.

As in the previous Chapters we use the interaction parameter $\lambda = \lambda_0(N - 1)$ to characterize the interaction strength. The quench is implemented as a sudden change of the interaction parameter from $\lambda = 0$ to $\lambda = 0.6$. Within the BH model this corresponds to a change from $U/J = 0$ to $U/J = 25.8$. The initial state is fully condensed $n^{(1)} = N$, and therefore superfluid. We note that the BH ground state at the final interaction strength is about 20% fragmented. In terms of the Lieb-Liniger parameter γ, the final interaction strength is such that $\gamma N^2 < 1.56$, which is in the 1D Gross-Pitaevskii regime according to the classification scheme in Sect. 2.11. The criteria (4.16) and (4.17) for the applicability of the BH model are well fulfilled and give $NU/e_{gap} \approx 1/8$ and $\mu/e_{gap} \approx 1/35$ for the state after the quench (with μ measured from the middle of the lowest band). The symmetry of the problem implies that $J(t)$ remains real, $\epsilon_L(t) = \epsilon_R(t) \equiv \epsilon(t)$, and $U_{LLLL}(t) = U_{RRRR}(t) \equiv U(t)$ at all times, which reduces the number of parameters in the TDBH model for this problem from six to three.

In Fig. 8.2 the natural orbital occupations $n_i^{(1)}$ of the TDBH model are shown together with those of the conventional BH model and those of the many-body Schrödinger equation obtained using MCTDHB. The MCTDHB results for the two largest natural orbital occupations using $M = 2$ and $M = 4$ orbitals coincide on the time-scale shown. For the remaining two natural orbital occupation numbers of the $M = 4$ computation we find $n_i^{(1)}/N < 4 \times 10^{-5}$ at all times. Thus, the results have converged, and the $M = 4$ results constitute numerically exact results of the many-body Schrödinger equation. The exact dynamics shows rapid oscillations of the fragmentation. The TDBH results follow the exact ones closely for many oscillations before deviating noticeably. These deviations prove that higher bands and/or more of the neglected terms have to be taken into account in the TDBH model in order to achieve quantitative agreement. The results of the conventional BH model deviate after much shorter times from the exact ones, which proves the necessity to use time-dependent Wannier functions.

While in real space hardly any dynamics is visible, a complex dynamics occurs in momentum space. Figure 8.3 shows the one-particle momentum distributions of the exact, the TDBH and the BH result at $t = 0$, where they all coincide and at later times where differences occur. Whenever the natural orbital occupations of the results are close also their one-particle momentum distributions $\rho(k; t) \equiv \rho^{(1)}(k|k; t)$, defined in Sect. 2.9, are very similar. The results of the BH model are obviously very different from those of the TDBH model and the exact ones, which are always nearby. The differences occur whenever the respective natural orbital occupations differ.

We have thereby shown that the use of time-dependent Wannier functions greatly improves the BH model. Of course the initial states of the BH model and the TDBH model have to be identical for this statement to be meaningful. The variational

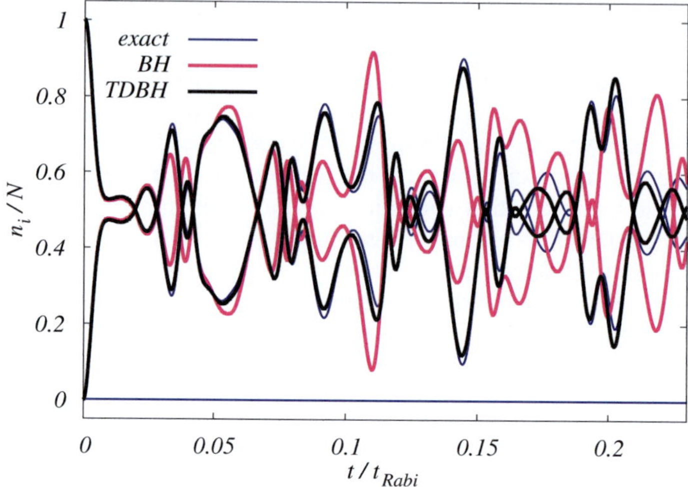

Fig. 8.2 Quantum quench dynamics in a double well. Shown are the natural orbital occupations following a quench from $\lambda = 0$ $(U/J = 0)$ to $\lambda = 0.6$ $(U/J = 25.8)$ as a function of time. The results of the BH model using time-independent (*magenta lines*), and time-dependent (*black lines*) Wannier functions are depicted together with those of the numerically exact solution of the many-body Schrödinger equation (*blue lines*). If time-independent Wannier functions are used the BH results deviate from the exact ones after short times. If time-dependent Wannier functions are used the BH results follow the exact ones closely. Starting from a fully condensed state the dynamics is complex and shows oscillations between partially and fully fragmented states

principle then ensures that the orbitals $\phi_L(x, t)$ and $\phi_R(x, t)$ are optimal at any point in time. Nevertheless, we note that even the TDBH model does not reproduce the exact many-body Schrödinger dynamics precisely.

The ratio U/J is the only relevant parameter of the BH model if time-independent Wannier functions are employed. In the present example the value for U/J of the BH model after the quench is $U/J = 25.8$ On the other hand the TDBH model allows the wave function to evolve more freely and $|itU(t)/J(t)$ varies in time. We are now in the position to assess to what extent the dynamics in the BH model is restricted by the assumption of time-independent Wannier functions. In Fig. 8.4 the ratio $U(t)/J(t)$ (top) and the time-dependent matrix elements $U(t)$ and $J(t)$ (below) of the TDBH model relative to their initial values are shown. Both $U(t)$ and $J(t)$ oscillate at a high frequency within bands. Note the time scale. The oscillations are caused by the shock imposed on the system through the quench. Just after the quench the density in each well expands due to the repulsive interaction between particles and hence $U(t)$ decreases, while $J(t)$ increases. $J(t)$ is very sensitive to these breathing oscillations. $J(t)$ varies over almost 25%, $U(t)$ over about 4%. As a result the ratio $U(t)/J(t)$ also varies. Over short periods of time $U(t)/J(t)$ varies between 26 and 20, a range of about 25%! Meanwhile, if time-independent Wannier functions are used the U and J are time-independent and the ratio $U/J = 25.8$ is constant at all times. This clearly demonstrates the need for time-dependent Wannier functions.

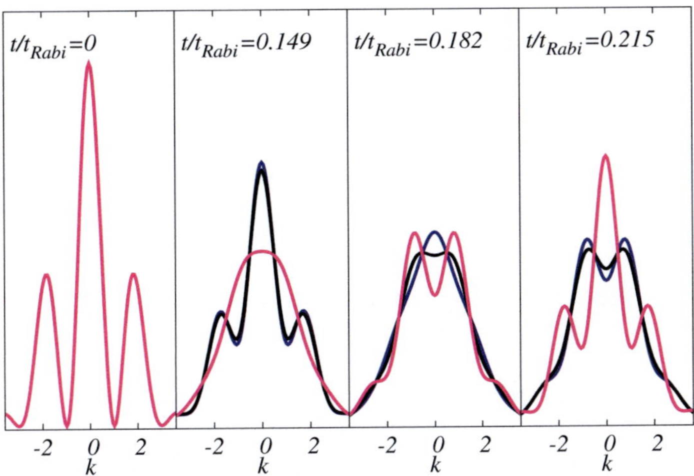

Fig. 8.3 Quench dynamics in a double well as in Fig. 8.2. Shown is the one-particle momentum distribution of the exact solution (*blue line*), the Bose–Hubbard model using time-dependent (*black line*) and time-independent (*magenta line*) Wannier functions at different times. The initial state is fully coherent and for some time the three results have similar momentum distributions (not shown). Differences between the results occur at later times whenever their respective natural orbital occupations differ, compare Fig. 8.2. The Bose–Hubbard result using time-dependent Wannier functions is always much closer to the exact one than the one using time-independent Wannier functions

8.4 Conclusions

We have generalized the concept of Wannier functions by allowing Wannier functions to be time-dependent. This additional degree of freedom can be exploited by deriving equations of motion from the time-dependent variational principle. The concept is general and can be applied to any lattice model that relies on Wannier functions. The additional cost of using time-dependent instead of time-independent Wannier functions in a lattice model is small compared to the cost of including higher bands. For any given initial state and any number of particles, lattice sites and bands, the use of time-dependent Wannier functions results in the variationally optimal answer within a given lattice model. Explicitly, we have derived equations of motion for the BH model with time-dependent Wannier functions. As a numerical example we have considered a quantum quench scenario. By comparison with the exact dynamics of the many-body Schrödinger equation we have shown that time-dependent Wannier functions greatly improve the BH model here. This is even more surprising as the advantages of time-dependent Wannier functions were not even nearly exploited in this example. The initial state was taken to be noninteracting and the dynamics was heavily restricted by the symmetry of the potential and initial state. Time-dependent Wannier functions are ideally suited for problems where these conditions are violated. In the context of the BH model it will therefore be particularly interesting to

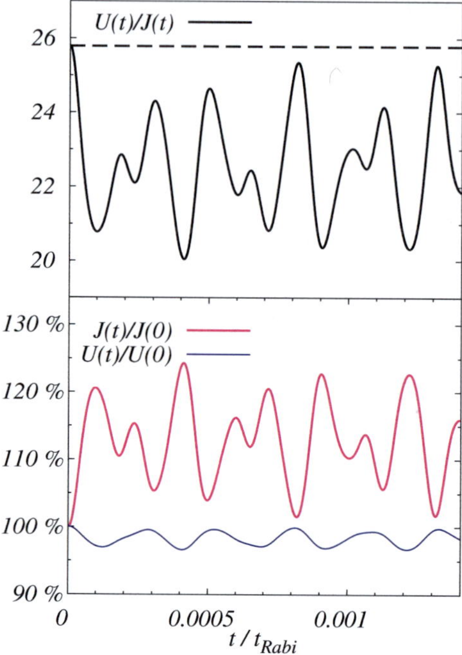

Fig. 8.4 Time-dependence of the hopping $J(t)$ and the on-site interaction matrix element $U(t)$ of the Bose–Hubbard model with time-dependent Wannier functions for the quench in Fig. 8.2. *Top*: the ratio $U(t)/J(t)$ oscillates rapidly with a large amplitude between about 26 and 20. Note the time scale. The ratio is always considerably smaller than its initial value (*dashed line*). *Bottom*: The ratios $J(t)/J(0)$ and $U(t)/U(0)$ oscillate in time. $J(t)$ is always larger, $U(t)$ always smaller than its initial value. The oscillations are due to a rapid expansion and contraction of the density following the quench. $J(t)$ varies over almost 25%, $U(t)$ over about 4% of its initial value. The respective BH results with time-independent Wannier functions are constant in time and are not shown

investigate the improvement that can be obtained for example in asymmetric double-well potentials or lattice potentials with an additional harmonic confinement. Similarly, the effects of disorder are often studied within the framework of lattice models. Such models will greatly benefit from time-dependent Wannier functions. Another interesting option is to study the quench dynamics in the double-well potential here, but using a different initial state. It will also be interesting to apply the concept of time-dependent Wannier functions to the fermionic Hubbard model and to investigate the improvement that can be obtained there.

References

1. K. Sakmann, A.I. Streltsov, O.E. Alon, L.S. Cederbaum, New J. Phys. **13**, 043003 (2011)
2. R. Pariser, R.G. Parr, J. Chem. Phys. **21**, 466 (1953)

3. R. Pariser, R.G. Parr, J. Chem. Phys. **21**, 767 (1953)
4. J.A. Pople, Trans. Faraday Soc. **49**, 1375 (1953)
5. J. Hubbard, Proc. Roy. Soc. (London), A **276**, 238 (1963)
6. M.C. Gutzwiller, Phys. Rev. Lett. **10**, 159 (1963)
7. J. Kanamori, Prog. Theor. Phys. **30**, 275 (1963)
8. M.P.A. Fisher, P.B. Weichman, G. Grinstein, D.S. Fisher, Phys. Rev. B **40**, 546 (1989)
9. K. Sakmann, A.I. Streltsov, O.E. Alon, L.S. Cederbaum, Phys. Rev. Lett. **103**, 220601 (2009)
10. K. Sakmann, A.I. Streltsov, O.E. Alon, L.S. Cederbaum, Phys. Rev A **82**, 013620 (2010)
11. V.W. Scarola, S. Das Sarma, Phys. Rev. Lett. **95**, 033003 (2005)
12. D.-S. Lühmann, K. Bongs, K. Sengstock, D. Pfannekuche, Phys. Rev. Lett. **101**, 050402 (2008)
13. D. Baillie , P. B. Blakie , Phys. Rev. A **80**, 033620 (2009)
14. A. Szabo, N.S. Ostlund, *Modern Quantum Chemistry*. (Dover Mineola, NY, 1996)
15. L.S. Cederbaum, O.E. Alon, A.I. Streltsov, Phys. Rev. A **73**, 043609 (2006)
16. A.I. Streltsov, O.E. Alon, L.S. Cederbaum, Phys. Rev. Lett. **99**, 030402 (2007)
17. O.E. Alon, A.I. Streltsov, L.S. Cederbaum, Phys. Rev. A **77**, 033613 (2008)
18. M.H. Beck, A. Jäckle, G.A. Worth, H.-D. Meyer, Phys. Rep. **324**, 1 (2000)
19. P. Würtz, T. Langen, T. Gericke, A. Koglbauer, H. Ott, Phys. Rev. Lett. **103**, 080404 (2009)
20. M. Greiner, O. Mandel, T. Esslinger, T. Hänsch, I. Bloch, Nature **415**, 39 (2002)
21. M. Greiner, O. Mandel, T. Hänsch, I. Bloch, Nature **419**, 51 (2002)
22. A.K. Tuchman, C. Orzel, A. Polkovnikov, M.A. Kasevich, Phys. Rev. A **74**, 051601(R) (2006)
23. T. Kinoshita, T. Wenger, D. Weiss, Nature **440**, 900 (2006)
24. C. Kollath, A.M. Läuchli, E. Altman, Phys. Rev. Lett. **98**, 180601 (2007)
25. M. Rigol, V. Dunjko, V. Yurovsky, M. Olshanii, Phys. Rev. Lett. **98**, 050405 (2007)
26. M. Rigol, V. Dunjko, M. Olshanii, Nature **452**, 854 (2008)
27. M. Rigol, Phys. Rev. Lett. **103**, 100403 (2009)

Chapter 9
Final Remarks and Outlook

In this thesis, trapped Bose-Einstein condensates were investigated based on the interacting many-body Schrödinger equation. Studies of this kind are (still) rare due to their computational complexity. Here, numerically *exact* results were obtained for the dynamics of up to one hundred identical bosons. This system size is unprecedented in literature. For the solution of the time-dependent many-body Schrödinger equation the recently developed MCTDHB method was used. Whenever appropriate, the results of the many-body Schrödinger equation were compared to the standard methods of the field, Gross–Pitaevskii theory and the Bose-Hubbard model. Thereby it was shown that the range of validity of these approximations is far more limited than what is commonly believed, and that no reliable validity criteria for these approximations exist to date. The true many-body physics of ultracold bosons turns out to be much richer than what can be anticipated based on Gross–Pitaevskii theory and the Bose-Hubbard model. This is even more so, as we have treated one of the conceptually simplest cases only, namely identical bosons trapped in a one-dimensional double-well potential. Therefore, the phenomena discovered here can only give a glimpse of what is to be expected in more complicated systems, e.g. optical lattices, higher dimensions, mixtures of bosons, boson-fermion mixtures and so on. Apart from the newly discovered physical phenomena, we have presented a conceptual innovation in this work: time-dependent Wannier functions. We have shown that lattice models can be greatly improved by letting the Wannier functions of a lattice model become time-dependent. The Bose-Hubbard model was used as an example to demonstrate this. However, the concept is general and can be applied to any lattice model that employs Wannier functions. It will be interesting to investigate how much other lattice models can be improved by using time-dependent Wannier functions. We have not included any treatment of higher dimensional systems in this thesis, but work is in progress. Another very interesting direction that we are currently pursuing is to map out explicitly the ranges of validity of Gross–Pitaevskii theory and the Bose-Hubbard model. Also here work is still in progress and results will soon be available.

K. Sakmann, *Many-Body Schrödinger Dynamics of Bose–Einstein Condensates*,
Springer Theses, DOI: 10.1007/978-3-642-22866-7_9,

Appendix A
IMEST-Algorithm for 1D, 2D and 3D Systems

A.1 Introduction

Often the two-body interaction potential that appears in the Hamiltonian of the many-body Schrödinger equation depends only on the distance between the particles. In the following we restrict the discussion to distance-dependent potentials. Note that, e.g. dipole–dipole interactions are not of this type unless all dipoles are aligned and confined to a plane. If a time-dependent single-particle basis is used, as for example in Gross–Pitaevskii (GP) theory or MCTDHB, the interaction matrix elements have to be evaluated at every time step of a propagation. In two and three spatial dimensions the evaluation of these integrals can become the performance limiting factor, especially when a long-range interaction potential is employed. For purely distance-dependent interactions the evaluation of interaction matrix elements can be greatly sped up by making explicit use of the functional form of the interaction potential and the fast Fourier transform (FFT). Also the evaluation of the matrix elements of the one-body Hamiltonian can be sped up by employing the FFT. The FFT is widely used technique in the propagation of nonlinear wave equations (known as split-step Fourier method) However, this does not seem to be the case for the evaluation of interaction matrix elements of the full many-body Hamiltonian. We therefore give here an algorithm that is ready to be implemented and compare its performance to a brute force approach.

A.2 Interaction Matrix Evaluation by Successive Transforms

We work in D dimensions and consider an interaction potential $W(|\mathbf{y}|)$, where $|\mathbf{y}| = |\mathbf{r} - \mathbf{r}'|$ is the distance between two particles located at $r = (x_1, \ldots, x_D)$ and $\mathbf{r}' = (x'_1, \ldots, x'_D)$. Consider the matrix element

K. Sakmann, *Many-Body Schrödinger Dynamics of Bose–Einstein Condensates*, Springer Theses, DOI: 10.1007/978-3-642-22866-7,

$$W_{ksql}(t) = \int\int d\mathbf{r}' d\mathbf{r} \phi_k^*(\mathbf{r},t)\phi_s^*(\mathbf{r}',t)W(\mathbf{r}-\mathbf{r}')\phi_q(\mathbf{r},t)\phi_l(\mathbf{r}',t). \tag{A.1}$$

It is possible to define a local time-dependent potential $W_{sl}(\mathbf{r},t)$ by

$$W_{sl}(\mathbf{r},t) = \int d\mathbf{r}' \phi_s(\mathbf{r}',t)^* W(\mathbf{r}-\mathbf{r}')\phi_l(\mathbf{r}',t) \tag{A.2}$$

For any given $W_{sl}(\mathbf{r},t)$ the evaluation of the matrix elements $W_{ksql}(t)$ then reduces to a single integral over $\mathbf{r}$:

$$W_{ksql}(t) = \int d\mathbf{r} \phi_k^*(\mathbf{r},t)W_{sl}(\mathbf{r},t)\phi_q(\mathbf{r},t). \tag{A.3}$$

Any two particle interaction potential of the form $W(\mathbf{r},\mathbf{r}') = W(\mathbf{r}-\mathbf{r}')$ has a Fourier transform of the following form:

$$W(\mathbf{r}-\mathbf{r}') = \frac{1}{(2\pi)^{D/2}} \int d\mathbf{k} \widehat{W}(\mathbf{k}) e^{i\mathbf{k}(\mathbf{r}-\mathbf{r}')} \tag{A.4}$$

$$\widehat{W}(\mathbf{k}) = \frac{1}{(2\pi)^{D/2}} \int d\mathbf{y} W(\mathbf{y}) e^{-i\mathbf{k}\mathbf{y}}. \tag{A.5}$$

Substituting Eq. A.4 in the expression Eq. A.2 results in

$$W_{sl}(\mathbf{r},t) = \int d\mathbf{r}' \phi_s(\mathbf{r}',t)^* \left[\frac{1}{(2\pi)^{D/2}} \int d\mathbf{k} \widehat{W}(\mathbf{k}) e^{i\mathbf{k}(\mathbf{r}-\mathbf{r}')}\right] \phi_l(\mathbf{r}',t) \tag{A.6}$$

$$W_{sl}(\mathbf{r},t) = \int d\mathbf{k} \left[\frac{1}{(2\pi)^{D/2}} \int d\mathbf{r}' \phi_s^*(\mathbf{r}',t)\phi_l(\mathbf{r}',t) e^{-i\mathbf{k}\mathbf{r}'}\right] \widehat{W}(\mathbf{k}) \mathbf{e}^{\mathbf{ikr}} \tag{A.7}$$

If we define the function

$$f_{sl}(\mathbf{r}',t) = \phi_s^*(\mathbf{r}',t)\phi_l(\mathbf{r}',t) \tag{A.8}$$

we see that the Fourier transform $\widehat{W}_{sl}(\mathbf{k},t)$ of $W_{sl}(\mathbf{r},t)$ can be expressed as

$$\widehat{W}_{sl}(\mathbf{k},t) = (2\pi)^{D/2} \widehat{f}_{sl}(\mathbf{k},\mathbf{t}) \widehat{\mathbf{W}}(\mathbf{k}) \tag{A.9}$$

$W_{sl}(\mathbf{r},t)$ is then given by the inverse Fourier transform of $W_{sl}(\mathbf{k},t)$:

$$W_{sl}(\mathbf{r},t) = \frac{1}{(2\pi)^{D/2}} \int d\mathbf{k} \widehat{W}_{sl}(\mathbf{k},t) e^{i\mathbf{k}\mathbf{r}}. \tag{A.10}$$

This way of evaluating $W_{sl}(\mathbf{r},t)$ requires three Fourier transformations: one Fourier transform of the interaction potential $\widehat{W}(\mathbf{k})$, one of the function $f_{sl}(\mathbf{r}',t)$ and one back transform of the function $\widehat{W}_{sl}(\mathbf{k},t)$. Numerically, Fourier transforms can be efficiently evaluated using FFT-algorithms. In practice only two Fourier transforms are necessary at each time step since the Fourier transform of the interaction potential $\widehat{W}(\mathbf{k})$ does not depend on time and has to be computed only once.

We now briefly outline how the evaluation of one-body matrix elements can be done efficiently using Fourier transforms. Consider the one-body part of the many-body Hamiltonian

$$h(\mathbf{r}) = -\frac{1}{2}\frac{\partial^2}{\partial \mathbf{r}^2} + V(\mathbf{r}). \tag{A.11}$$

A matrix element $h_{kq}(t)$ as defined in Eq. 2.10 can then be evaluated as follows.

$$h_{kq}(t) = \int d\mathbf{r}\left(\phi_k^*(\mathbf{r},t) - \frac{1}{2}\frac{\partial^2}{\partial \mathbf{r}^2}\phi_q(\mathbf{r},t) + \phi_k^*(\mathbf{r},t)V(\mathbf{r})\phi_q(\mathbf{r},t)\right). \tag{A.12}$$

The second part of the integral is easily evaluated in real space and will not be discussed any further. There are many different ways to evaluate the first part of the integral that contains the Laplacian in D dimensions. Here we choose to transform the orbital ϕ_q to momentum space, where the representation of the Laplacian becomes a diagonal matrix:

$$\int d\mathbf{r}\phi_k^*(\mathbf{r},t)\frac{\partial^2}{\partial \mathbf{r}^2}\phi_q(\mathbf{r},t) = \frac{1}{(2\pi)^{D/2}}\int d\mathbf{r}\int d\mathbf{k}\phi_k^*(\mathbf{r},t)(-\mathbf{k}^2)\phi_q(\mathbf{k},t)e^{i\mathbf{k}\mathbf{r}} \tag{A.13}$$

Defining $\Phi_q(\mathbf{k},t) \equiv -\mathbf{k}^2\phi_q(\mathbf{k},t)$, substituting

$$\Phi_q(\mathbf{k},t) = \frac{1}{(2\pi)^{D/2}}\int d\mathbf{r}'\Phi_q(\mathbf{r},t)e^{-i\mathbf{k}\mathbf{r}} \tag{A.14}$$

into (A.13) and using the identity $\delta(\mathbf{r}-\mathbf{r}') = (2\pi)^D \int d\mathbf{k}e^{i\mathbf{k}(\mathbf{r}-\mathbf{r}')}$ we find

$$\int d\mathbf{r}\phi_k^*(\mathbf{r},t)\frac{\partial^2}{\partial \mathbf{r}^2}\phi_q(\mathbf{r},t) = \int d\mathbf{r}\phi_k^*(\mathbf{r},t)\Phi_q(\mathbf{r},t). \tag{A.15}$$

This way of evaluating the action of the Laplacian requires two Fourier transforms, which can be efficiently implemented using the FFT algorithm.

A.3 Theory of the IMEST-Algorithm in Finite, Discrete Space

Let us assume that for any fixed $\mathbf{r}$ the integrand in Eq. A.2 decays sufficiently rapidly with $|\mathbf{r}'-\mathbf{r}|$ to approximate the integral in Eq. A.2 by an integral over the box $V = [a_1, b_1] \times \cdots \times [a_D, b_D]$. Without loss of generality we can assume that $a_j = 0$ and $b_j = L_j$ for $j = 1, \cdots, D$. In the following we therefore work with

$$V = [0, L_1] \times \cdots \times [0, L_D]. \tag{A.16}$$

If $\mathbf{r}$ and $\mathbf{r}'$ are restricted to V then $\mathbf{y} = \mathbf{r} - \mathbf{r}'$ is restricted to $\overline{V} = [-L_1, L_1] \times \cdots \times [-L_D, L_D]$. The orbitals $\phi_s(\mathbf{r})$, $\phi_l(\mathbf{r})$ and the interaction potential $W(\mathbf{y})$ are therefore not defined on the same domain if the integral in Eq. A.2 is taken over V only. However, if $W(\mathbf{y})$ decays sufficiently rapidly at large distances $|\mathbf{y}| \to \infty$, the Fourier transform $\widehat{W}(\mathbf{k})$ of $W(\mathbf{r})$ can be approximated by

$$\widehat{W}(\mathbf{k}) = \frac{1}{(2\pi)^{D/2}} \int_{-L_1/2}^{L_1/2} \cdots \int_{-L_D/2}^{L_D/2} d\mathbf{y} W(\mathbf{y}) e^{-i\mathbf{k}\mathbf{y}}. \tag{A.17}$$

If we define the function $w(\mathbf{y}) = W(y_1 - L_1/2, \ldots, y_D - L_D/2)$, then Eq. A.17 can be expressed as an integral over V:

$$\widehat{W}(\mathbf{k}) = \int_V d\mathbf{y} w(\mathbf{y}) e^{-i\mathbf{k}\mathbf{y}}. \tag{A.18}$$

The Fourier transform $\widehat{W}(\mathbf{k})$ is then related to that of $w(\mathbf{r})$ by

$$\widehat{W}(\mathbf{k}) = \widehat{w}(\mathbf{k}) e^{+i \sum_{j=1}^{D} k_j \frac{L_j}{2}}. \tag{A.19}$$

We now make the transition to discrete space. For $j = 1, \cdots, D$ we define n_j grid points along each coordinate axis x_j by

$$x_{\alpha_j} = \alpha_j \Delta x_j, \alpha_j = 0, \ldots, n_j - 1, \tag{A.20}$$

with $\Delta x_j = L_j/n_j$. For simplicity we assume that all n_j are even. Similarly, we define a grid in momentum space by

$$k_{\gamma_j} = \gamma_j \Delta k_j, \gamma_j = -n_j/2, \ldots, n_j/2 - 1, \tag{A.21}$$

for $j = 1, \cdots, D$, where $\Delta k_j = 2\pi/L_j$. If $z(\mathbf{r})$ is a function of $\mathbf{r}$ we write

$$z_{\alpha_1\alpha_2\ldots\alpha_D} = z(x_{\alpha_1}, \ldots, x_{\alpha_D}). \tag{A.22}$$

We then collect all points $z_{\alpha_1\alpha_2\ldots\alpha_D}$ in a single vector $\mathbf{z}$. We then define the elements $\widetilde{z}_{\beta_1\beta_2\ldots\beta_D}$ of the D-dimensional forward discrete Fourier transform (DFT), $\mathcal{F}^-$, of $\mathbf{z}$ by

$$\begin{aligned} \widetilde{z}_{\beta_1\beta_2\ldots\beta_D} &= \frac{1}{\sqrt{n_1 n_2 \ldots n_D}} \sum_{\alpha_1=0}^{n_1-1} \cdots \sum_{\alpha_D=0}^{n_D-1} z_{\alpha_1\alpha_2\ldots\alpha_D} e^{-2\pi i \alpha_1 \beta_1/n_1} \ldots e^{-2\pi i \alpha_D \beta_D/n_D} \\ &= \mathcal{F}^-_{\beta_1\beta_2\ldots\beta_D}(\mathbf{z}), \beta_j = 0, \ldots, n_j - 1, j = 1, \ldots, D. \end{aligned} \tag{A.23}$$

With this definition of the forward DFT the elements of the backward DFT, $\mathcal{F}^+$, are given by

$$\begin{aligned} z_{\alpha_1\alpha_2\ldots\alpha_D} &= \frac{1}{\sqrt{n_1 n_2 \ldots n_D}} \sum_{\beta_1=0}^{n_1-1} \cdots \sum_{\beta_D=0}^{n_D-1} \widetilde{z}_{\beta_1\beta_2\ldots\beta_D} e^{+2\pi i \alpha_1 \beta_1/n_1} \ldots e^{+2\pi i \alpha_D \beta_D/n_D} \\ &= \mathcal{F}^{+}_{\alpha_1\alpha_2\ldots\alpha_D}(\widetilde{\mathbf{z}}), \alpha_j = 0,\ldots,n_j-1, j=1,\ldots,D. \end{aligned} \tag{A.24}$$

The elements $\widetilde{z}_{\beta_1\beta_2\ldots\beta_D}$ are periodic in all indices $\widetilde{z}_{\beta_1\ldots\beta_j\ldots\beta_D} = \widetilde{z}_{\beta_1\ldots\beta_j+n_j\ldots\beta_D}$ for $j = 1,\ldots,D$. The value of the Fourier transform $\widehat{w}(\mathbf{k})$ at the grid point $\mathbf{k} = (k_{\gamma_1},\ldots,k_{\gamma_D})$ is then approximated by its DFT value

$$\widehat{w}(\beta_1\Delta k_1,\ldots,\beta_D\Delta k_D) \approx \widehat{w}_{\beta_1\beta_2\ldots\beta_D} = \Delta V \frac{\sqrt{n}}{(2\pi)^{D/2}} \widetilde{w}_{\beta_1\beta_2\ldots\beta_D}, \tag{A.25}$$

where $\Delta V = \Delta x_1 \Delta x_2 \ldots \Delta x_D$ and $n = n_1 n_2 \ldots n_D$. Due to the periodicity of $\widetilde{z}_{\beta_1\ldots\beta_j\ldots\beta_D}$ the first $n_j/2$ elements of the index β_j refer to positive momenta $k_{\beta_j} = 0,\ldots,(n_j/2-1)\Delta k_j$, and the following $n_j/2$ elements refer to negative momenta $k_{\beta_j} = -(n_j/2)\Delta k_j,\ldots,-\Delta k_j$. By defining $v = \beta_1 + \beta_2 + \ldots + \beta_D$ and using Eqs. A.19 and A.25 we find for the DFT approximation of $\widehat{W}(\mathbf{k})$

$$\widehat{W}(\beta_1\Delta k_1\ldots\beta_D\Delta k_D) \approx \widehat{W}_{\beta_1\beta_2\ldots\beta_D} = \Delta V \frac{\sqrt{n}}{(2\pi)^{D/2}} \widetilde{w}_{\beta_1\beta_2\ldots\beta_D}(-1)^v \tag{A.26}$$

at $\mathbf{k} = (k_{\gamma_1},\ldots,k_{\gamma_D})$. We collect the values of the function $f_{sl}(\mathbf{r},t)$ at the grid points $\mathbf{r} = (x_{\alpha_1},\cdots,x_{\alpha_D})$ in a single vector $\mathbf{f}(t)$ with elements $f_{sl\alpha_1\alpha_2\cdots\alpha_D}(t)$. Analogous to Eq. A.25 one finds

$$\widehat{f}_{sl\beta_1\beta_2\cdots\beta_D}(t) = \Delta V \frac{\sqrt{n}}{(2\pi)^{D/2}} \widetilde{f}_{sl\beta_1\beta_2\cdots\beta_D}(t) \tag{A.27}$$

$$\begin{aligned} \widehat{W}_{sl\beta_1\beta_2\cdots\beta_D}(t) &= (2\pi)^{D/2} \widehat{f}_{sl\beta_1\beta_2\cdots\beta_D}(t) \widehat{W}_{\beta_1\beta_2\cdots\beta_D} \\ &= (\Delta V)^2 \frac{n}{(2\pi)^{D/2}} \left[(-1)^v \widetilde{f}_{sl\beta_1\beta_2\cdots\beta_D}(t) \widetilde{w}_{\beta_1\beta_2\cdots\beta_D}\right] \end{aligned} \tag{A.28}$$

as DFT approximations to the Fourier transforms $\widehat{f}_{sl}(\mathbf{k},t)$ and $\widehat{W}_{sl}(\mathbf{k},t)$ at $\mathbf{k} = (k_{\gamma_1},\ldots,k_{\gamma_D})$. We collect all values $\widehat{f}_{sl\beta_1\beta_2\cdots\beta_D}(t)$ in a vector $\widehat{\mathbf{f}}_{sl}(t)$ and all values $\widehat{W}_{sl\beta_1\beta_2\cdots\beta_D}(t)$ in a vector $\widehat{\mathbf{W}}_{sl}(t)$. It remains to transform $\widehat{\mathbf{W}}_{sl}(t)$ back to real space. The DFT value of $W_{sl}(\mathbf{r},t)$ at $\mathbf{r} = (x_{\alpha_1},\ldots,x_{\alpha_D})$ is then given by

$$W_{sl\alpha_1\alpha_2\ldots\alpha_D}(t) = \Delta k_1 \cdots \Delta k_D \frac{\sqrt{n}}{(2\pi)^{D/2}} \mathcal{F}^{+}_{\alpha_1\alpha_2\ldots\alpha_D}(\widehat{\mathbf{W}}_{sl}(t)) \tag{A.29}$$

We collect all elements $W_{sl\alpha_1\alpha_2\ldots\alpha_D}(t)$ in a vector $\mathbf{W}_{sl}(t)$ and define the vector $\widetilde{\mathbf{v}}$ by

$$\widetilde{v}_{\beta_1\beta_2\cdots\beta_D} = \sqrt{n}(-1)^v \widetilde{w}_{\beta_1\beta_2\cdots\beta_D} \tag{A.30}$$

The elements $\widetilde{W}_{sl\beta_1\beta_2\ldots\beta_D}(t)$ of the DFT of W_{sl} are then given by

$$\widetilde{W}_{sl\beta_1\beta_2\ldots\beta_D} = \widetilde{v}_{\beta_1\beta_2\ldots\beta_D}\widetilde{f}_{sl\beta_1\beta_2\ldots\beta_D}(t)\Delta V \tag{A.31}$$

and the vector $\mathbf{W}_{sl}(t)$ is obtained by applying the backward DFT $\mathcal{F}^+$ to $\widetilde{\mathbf{W}}_{sl}(t)$

$$\mathbf{W}_{sl}(t) = \mathcal{F}^+(\widetilde{\mathbf{W}}_{sl}(t)). \tag{A.32}$$

The IMEST-algorithm to evaluate $\mathbf{W}_{sl}(t)$ and $W_{ksql}(t)$ is summarized in Algorithm 1.

A.4 Performance Test of the IMEST-Algorithm

In this section we present a performance analysis of the IMEST-algorithm in one, two and three dimensions. The accuracy and the speed of the IMEST-algorithm are discussed. As a benchmark we choose the computation of the vector $\mathbf{W}_{sl}(t)$. We choose to work on a grid with equal numbers $n_j = n^{1/D}$ of grid points along each coordinate. For comparison we compute $\mathbf{W}_{sl}(t)$ also by evaluating the integral in Eq. A.2 at each grid point $\mathbf{r}$ separately. We refer to the latter method as brute force (BF) evaluation.

The complexity of the IMEST-algorithm is expected to be about $\mathcal{O}(n\log_n)$ since two FFTs are performed. The brute force algorithm on the other hand is expected to have a complexity of $\mathcal{O}(n^2)$ since for each of the n grid points one integral over n grid points has to be evaluated. Figure A.1 shows the average CPU-time needed for the evaluation of the vector $\mathbf{W}_{sl}(t)$ as a function of n_j on an AMD dual-core Opteron processor at 2.6 GHz. The times are obtained by averaging over

Algorithm 1 IMEST-Algorithm in D-dimensions

Evaluate $\mathbf{w} = (w_0, \cdots, w_{n-1})$
Compute$\widetilde{\mathbf{w}} = \mathcal{F}^-(\mathbf{w})$
for $m = 0, \ldots, n-1$ **do**
evaluate $\widetilde{v}_m = \sqrt{n}(-1)^\nu \widetilde{w}_m$ with $\nu = \beta_1 + \ldots + \beta_D$
end for
Store $\widetilde{\mathbf{v}}$ permanently
Choose orbitals ϕ_s and ϕ_l
for $m = 0, \ldots, n-1$ **do**
evaluate $f_{slm}(t) = \phi_s(x_m, t)^* \phi_l(x_m, t)$
end for
Compute $\widetilde{\mathbf{f}}_{sl} = \mathcal{F}^-(\mathbf{f}_{sl}(t))$
for $m = 0, \ldots, n-1$ **do**
evaluate $\widetilde{u}_{slm}(t) = \Delta V \widetilde{f}_{slm}(t)\widetilde{v}_m$
end for
Compute $\mathbf{W}_{sl}(t) = \mathcal{F}^+(\mathbf{u}_{sl}(t))$
Choose orbitals ϕ_k and ϕ_q
Compute $W_{ksql}(t) = \Delta V \sum_{m=0}^{n-1} \phi_k^*(x_m) W_{slm}(t) \phi_q(x_m, t)$

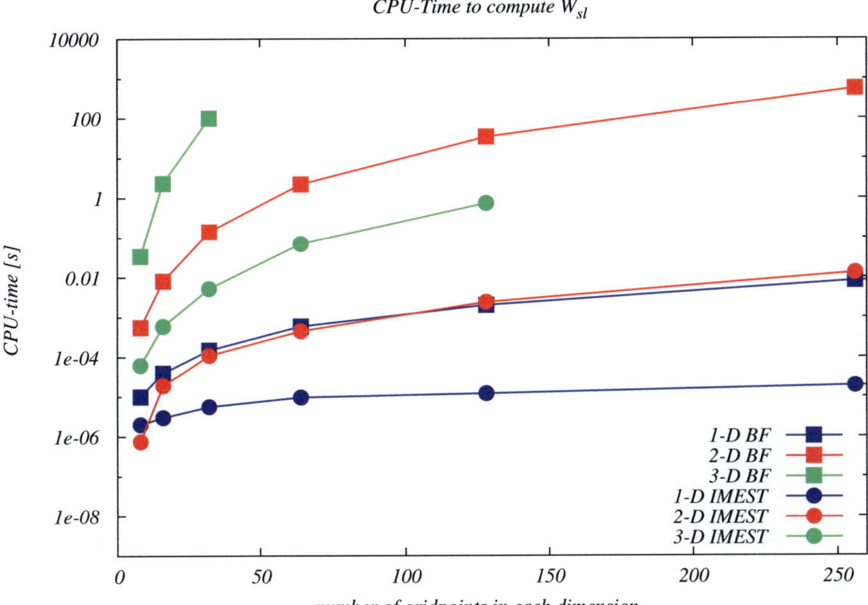

Fig. A.1 Performance of the IMEST-algorithm and the brute force algorithm. Shown is the average CPU-time needed for the evaluation of $\mathbf{W}_{sl}$ as a function of the number of grid points in each direction $n_j, j = 1, \ldots, D$ (see text for details) and different dimensions $D = 1, 2, 3$. The times are obtained by averaging over fifty evaluations of $\mathbf{W}_{sl}$. The CPU-times of the IMEST-algorithm (*lines with circles*) are always below the CPU-times of the brute force algorithm (*lines with squares*). Even in 1D the IMEST-algorithm is orders of magnitude faster than the brute force algorithm. The larger the dimension the greater is the difference between the CPU-times of the two algorithms. The CPU-time needed by the IMEST algorithm in 2D is about the same as that of the brute force algorithm in 1D up to about $n_j = 256$

fifty evaluations of $\mathbf{W}_{sl}$. Even in $D = 1$ dimension with $n_j = 8$ grid points the IMEST-algorithm is faster than the brute force algorithm. For larger grid sizes the IMEST algorithm is orders of magnitude faster. The advantages of using the IMEST-algorithm is even more striking if $D > 1$ since the total number of grid points increases exponentially with D. In a quantum dynamics computation the vector $\mathbf{W}_{sl}(t)$ has to be evaluated at every time step. It is therefore possible to estimate from Fig. A.1 the limitations that the evaluation of $\mathbf{W}_{sl}(t)$ puts on the grid size and dimensions that are feasible. If a maximally feasible CPU-time of $t_{max} = 1s$ is chosen for the evaluation of $\mathbf{W}_{sl}(t)$ then computations $D = 2$ and $D = 3$ are not feasible for $n_j \geq 32$ grid points, if the brute force algorithm is used. The IMEST-algorithm makes such computations feasible. In $D = 2$ dimensions the cost of evaluating $\mathbf{W}_{sl}$ via the IMEST-algorithm is roughly the same as the cost of evaluating $\mathbf{W}_{sl}$ in $D = 1$ via the brute force algorithm.

Appendix B
p-Particle Momentum Distribution

It can be shown that the p-particle momentum distribution at large momenta is dominated by contributions of $\rho^{(p)}(\mathbf{r}_1,\ldots,\mathbf{r}_p|\mathbf{r}'_1,\ldots,\mathbf{r}'_p;t)$ close to the diagonal, i.e. $\mathbf{r}_i \approx \mathbf{r}'_i$ for $i=1,\ldots,p$. Similarly, the p-particle distribution at low momenta is dominated by the behavior of $\rho^{(p)}(\mathbf{r}_1,\ldots,\mathbf{r}_p|\mathbf{r}'_1,\ldots,\mathbf{r}'_p;t)$ on the off-diagonal at large distances between $\mathbf{r}_i$ and $\mathbf{r}'_i$.

The p-particle RDM is related to the p-particle momentum distribution by

$$\begin{aligned}\rho^{(p)}(\mathbf{k}_1,\ldots,\mathbf{k}_p|\mathbf{k}_1,\ldots,\mathbf{k}_p;t) = \frac{1}{(2\pi)^{Dp}}\int d^p\mathbf{r}d^p\mathbf{r}'e^{-i\sum_{l=1}^{p}\mathbf{k}_l(\mathbf{r}_l-\mathbf{r}'_l)} \\ \times\,\rho^{(p)}(\mathbf{r}_1,\ldots,\mathbf{r}_p|\mathbf{r}'_1,\ldots,\mathbf{r}'_p;t).\end{aligned} \tag{B.1}$$

The change of variables $\mathbf{R}_i=\frac{\mathbf{r}_i+\mathbf{r}'_i}{2}$, $\mathbf{s}_i=\mathbf{r}_i-\mathbf{r}'_i$ for $i=1,\ldots,p$ in Eq. B.1 leads to

$$\rho^{(p)}(\mathbf{k}_1,\ldots,\mathbf{k}_p|\mathbf{k}_1,\ldots,\mathbf{k}_p;t) = \int d^p\mathbf{s}e^{-i\sum_{l=1}^{p}\mathbf{k}_l\mathbf{s}_l}\gamma^{(p)}(\mathbf{s}_1,\ldots,\mathbf{s}_p;t), \tag{B.2}$$

where

$$\gamma^{(p)}(\mathbf{s}_1,\ldots,\mathbf{s}_p;t) = \int d^p\mathbf{R}\rho^{(p)}(\mathbf{R}_1+\frac{\mathbf{s}_1}{2},\ldots,\mathbf{R}_p+\frac{\mathbf{s}_p}{2}|\mathbf{R}_1-\frac{\mathbf{s}_1}{2},\ldots,\mathbf{R}_p-\frac{\mathbf{s}_p}{2};t) \tag{B.3}$$

is the average of the value of $\rho^{(p)}(\mathbf{r}_1,\ldots,\mathbf{r}_p|\mathbf{r}'_1,\ldots,\mathbf{r}'_p;t)$ at distances $\mathbf{s}_i$ between $\mathbf{r}_i$ and $\mathbf{r}'_i$. From Eqs. B.2, B.3 it is clear thlat the p-particle momentum distribution at large momenta is determined by the behavior of $\rho^{(p)}(\mathbf{r}_1,\ldots,\mathbf{r}_p|\mathbf{r}'_1,\ldots,\mathbf{r}'_p;t)$ at short distances, whereas at low momenta the off-diagonal long range behavior of $\rho^{(p)}(\mathbf{r}_1,\ldots,\mathbf{r}_p|\mathbf{r}'_1,\ldots,\mathbf{r}'_p;t)$ contributes the major part.

K. Sakmann, *Many-Body Schrödinger Dynamics of Bose–Einstein Condensates*,
Springer Theses, DOI: 10.1007/978-3-642-22866-7,

Appendix C
Matrix Elements of the 1st and 2nd Order RDMs

In this Appendix we list explicitly expressions for the first and second order RDMs of multi-configurational bosonic [1] wave functions $|\Psi(t)\rangle = \sum_{\vec{n}} C_{\vec{n}}(t)$ $|n_1, n_2, \ldots, n_M; t\rangle$. We also need a shorthand notation for a reference configuration and relevant excited configurations. Let the reference configuration be denoted by

$$|\vec{n}; t\rangle = \left|n_1, \ldots, n_k, \ldots, n_s, \ldots, n_l, \ldots, n_q, \ldots, n_M; t\right\rangle. \tag{C.1}$$

Then, the configuration denoted by

$$\left|\vec{n}_k^q; t\right\rangle = \left|n_1, \ldots, n_k - 1, \ldots, n_s, \ldots, n_l, \ldots, n_q + 1, \ldots, n_M; t\right\rangle \tag{C.2}$$

differs from $|\vec{n}; t\rangle$ by an excitation of one particle from the k-th to the q-th orbital;

$$\left|\vec{n}_{kk}^{ll}; t\right\rangle = \left|n_1, \ldots, n_k - 2, \ldots, n_s, \ldots, n_l + 2, \ldots, n_q, \ldots, n_M; t\right\rangle \tag{C.3}$$

represents excitations of two particles from the k-th to the l-th orbital;

$$\left|\vec{n}_{kk}^{ql}; t\right\rangle = \left|n_1, \ldots, n_k - 2, \ldots, n_s, \ldots, n_l + 1, \ldots, n_q + 1, \ldots, n_M; t\right\rangle \tag{C.4}$$

represents excitations of two particles from the k-th orbital, one to the q-th and the other to the l-th orbital; and

$$\left|\vec{n}_{ks}^{ql}; t\right\rangle = \left|n_1, \ldots, n_k - 1, \ldots, n_s - 1, \ldots, n_l + 1, \ldots, n_q + 1, \ldots, n_M; t\right\rangle \tag{C.5}$$

represents excitations of two particles, one from the k-th to the q-th orbital and a second particle from the s-th to the l-th orbital. Note that we do not utilize excitation operators to define the excited configurations atop the reference configuration. Rather, it is convenient for our needs to employ a nomenclature in which the *same* ordering of the orbitals $\phi_1, \phi_2, \ldots, \phi_M$ as in Eq. 2.15 is kept in *all*

K. Sakmann, *Many-Body Schrödinger Dynamics of Bose–Einstein Condensates*, Springer Theses, DOI: 10.1007/978-3-642-22866-7,

configurations. In this nomenclature the following states are equivalent: $\left|\vec{n}^{ql}_{ks};t\right\rangle \equiv \left|\vec{n}^{ql}_{sk};t\right\rangle \equiv \left|\vec{n}^{lq}_{sk};t\right\rangle \equiv \left|\vec{n}^{lq}_{ks};t\right\rangle$.

With these observations and notations, the elements of the first order RDM $\rho^{(1)}(\mathbf{r}_1|\mathbf{r}'_1;t)$ of the multi-configurational ansatz $|\Psi(t)\rangle = \sum_{\vec{n}} C_{\vec{n}}(t)|\vec{n};t\rangle$ are:

$$\rho_{kk}(t) = \sum_{\vec{n}} C^*_{\vec{n}} C_{\vec{n}} n_k, \rho_{kq}(t) \tag{C.6}$$

From the hermiticity of the one-body operator we readily have $\rho_{kq}(t) = \rho^*_{qk}(t)$.

The elements of the second order RDM,$\rho^{(2)}(\mathbf{r}_1, \mathbf{r}_2|\mathbf{r}'_1, \mathbf{r}'_2;t)$, given the multi-configurational ansatz $|\Psi(t)\rangle = \sum_{\vec{n}} C_{\vec{n}}(t)|\vec{n};t\rangle$ are:

$$\begin{aligned}
\rho_{kkkk}(t) &= \sum_{\vec{n}} C^*_{\vec{n}} C_{\vec{n}} n_k [n_k - 1], \\
\rho_{kkkq}(t) &= \sum_{\vec{n}} C^*_{\vec{n}} C_{\vec{n}^q_k} [n_k - 1]\sqrt{n_k[n_q + 1]},, \\
\rho_{kkll}(t) &= \sum_{\vec{n}} C^*_{\vec{n}} C_{\vec{n}^{ll}_{kk}} \sqrt{[n_k - 1]n_k[n_l + 1][n_l + 2]}, \\
\rho_{kssk}(t) &= \sum_{\vec{n}} C^*_{\vec{n}} C_{\vec{n}} n_k n_s, \\
\rho_{kklq}(t) &= \sum_{\vec{n}} C^*_{\vec{n}} C_{\vec{n}^{ql}_{kk}} \sqrt{[n_k - 1]n_k[n_l + 1][n_q + 1]}, \\
\rho_{kssq}(t) &= \sum_{\vec{n}} C^*_{\vec{n}} C_{\vec{n}^q_k} n_s \sqrt{n_k[n_q + 1]}, \\
\rho_{kslq}(t) &= \sum_{\vec{n}} C^*_{\vec{n}} C_{\vec{n}^{ql}_{ks}} \sqrt{n_k n_s [n_l + 1][n_q + 1]},
\end{aligned} \tag{C.7}$$

where it is understood that different indices k, s, q, l do not have the same value. All other non-vanishing matrix elements can be computed due to the symmetries of the two-body operator, $\rho_{kslq} = \rho_{sklq} = \rho_{skql} = \rho_{ksql}$ and its hermiticity, $\rho^*_{kslq} = \rho_{lqks}$.

Appendix D
Matrix Elements of the Operators $\hat{h} - i\frac{\partial}{\partial t}$ and $\hat{W}$

In this Appendix we list rules for evaluating matrix elements with permanents [1].

We use the conventions introduced in Appendix C and the time-dependent matrix elements of the one- and two-body operators with respect to the orbitals given in Eqs. 2.10 and 3.5 of the main text. The non-vanishing matrix elements of the one-body operator $\hat{h} - i\frac{\partial}{\partial t}$ follow from:

$$\left\langle \vec{n};t \middle| \hat{h} - i\frac{\partial}{\partial t} \middle| \vec{n};t \right\rangle = \sum_{l=1}^{M} n_l \left[h_{ll} - \left(i\frac{\partial}{\partial t} \right)_{ll} \right], \left\langle \vec{n};t \middle| \hat{h} - i\frac{\partial}{\partial t} \middle| \vec{n}_k^q;t \right\rangle \tag{D.1}$$

and the fact that the one-body operator $\hat{h} - i\frac{\partial}{\partial t}$ is self-adjoint,

$$\left\langle \vec{n};t \middle| \hat{h} - i\frac{\partial}{\partial t} \middle| \vec{n}';t \right\rangle = \left\langle \vec{n}';t \middle| \hat{h} - i\frac{\partial}{\partial t} \middle| \vec{n};t \right\rangle^*. \tag{D.2}$$

The non-vanishing matrix elements of the two-body operator $\hat{W}$ follow from:

$$\begin{aligned}
\langle \vec{n};t | \hat{W} | \vec{n};t \rangle &= \frac{1}{2}\sum_{j=1}^{M} n_j \left([n_j - 1] W_{jjjj} + \sum_{\{i \neq j\}=1}^{M} n_i W_{ji\{ij\}} \right), \\
\langle \vec{n};t | \hat{W} | \vec{n}_k^q;t \rangle &= \sqrt{n_k[n_q+1]} \left([n_k - 1] W_{kkkq} + n_q W_{kqqq} + \sum_{\{i \neq k,q\}=1}^{M} n_i W_{ki\{iq\}} \right), \\
\langle \vec{n};t | \hat{W} | \vec{n}_{kk}^{ll};t \rangle &= \frac{1}{2}\sqrt{[n_k - 1] n_k [n_l+1][n_l+2]} W_{kkll}, \\
\left\langle \vec{n};t | \hat{W} | \vec{n}_{kk}^{ql}; \right\rangle &= \sqrt{[n_k - 1] n_k [n_l+1][n_q+1]} W_{kklq}, \\
\left\langle \vec{n};t | \hat{W} | \vec{n}_{ks}^{ql};t \right\rangle &= \sqrt{n_k n_s [n_l+1][n_q+1]} W_{ks\{lq\}}
\end{aligned} \tag{D.3}$$

K. Sakmann, *Many-Body Schrödinger Dynamics of Bose–Einstein Condensates*,
Springer Theses, DOI: 10.1007/978-3-642-22866-7,

where it is understood that different indices k, s, q, l do not have the same value and

$$W_{ks\{lq\}} = W_{kslq} + W_{ksql}. \tag{D.4}$$

The fact that the two-body operator $\hat{W}$ is self-adjoint entails $\langle \vec{n}; t | \hat{W} | \vec{n}'; t \rangle = \langle \vec{n}'; t | \hat{W} | \vec{n}; t \rangle^*$.

Appendix E
Best Mean-Field

The exact many-body wave function of a bosonic system of N particles can always be expanded in an infinite weighted sum over any complete set of permanents of N particles. We restrict ourselves to stationary states here. In mean-field theory the exact many-body wave function is approximated by a single permanent, see Sect. 2.4. This single permanent is built from a number $M \leq N$ of orthogonal orbitals in which the N bosons reside. In the field of Bose-Einstein condensates one particular mean-field, the GP mean-field, has proven to be very successful. In analogy to non-interacting BECs, in GP theory it is assumed that the many-body wave function is given by a single permanent in which all particles reside in *one* orbital, i.e. $M = 1$. A minimization of the energy functional with the GP *ansatz* wave function leads to the famous Gross–Pitaevskii equation [2, 3, 4, 5]. The solution of the GP equation yields the single orbital from which the GP mean-field permanent is constructed.

However, it has been shown [6–10] that the GP mean-field is not always the energetically-lowest mean-field solution. The assumption that *all* particles occupy the same orbital is too restrictive. Especially in multi-well trapping geometries the energetically-lowest mean-field solution can be fragmented [6–10], see also Sect. 2.5.

In order to obtain the energetically-lowest mean-field solution, it is necessary that the ansatz for the wave function is of the most general mean-field form. Due to the variational principle, the minimization of the respective energy functional with respect to all parameters of the ansatz wave function will then give the *best* solution within mean-field theory. It is therefore legitimate to call this mean-field solution the best mean field (BMF). A procedure to obtain the BMF solution numerically has been developed recently [6–10].

In the best mean-field approach the ansatz for the wave function $|\Psi\rangle$ is taken as a single permanent of N bosons distributed over M time-independent orthonormal orbitals $\phi_k(\mathbf{r})$:

$$|\Psi\rangle = |n_1, n_2, \ldots, n_M\rangle. \tag{E.1}$$

K. Sakmann, *Many-Body Schrödinger Dynamics of Bose–Einstein Condensates*, Springer Theses, DOI: 10.1007/978-3-642-22866-7,

Using this ansatz for the wave function, the energy functional is minimized by a variation over the number of orbitals M, the occupation numbers n_i and the orbitals $\phi_k(\mathbf{r})$ themselves [6, 10]. The variation leads to a set of coupled non-linear equations that have to be solved to obtain the BMF solution. Thereby, the energetically most favourable permanent is selected to approximate the true many-body wave function. The GP mean-field is contained in the BMF ansatz as can be seen by restricting the number of orbitals to $M = 1$.

Appendix F
On the Importance of Time-Dependent Basis Sets

F.1 Introduction

In this thesis we have made extensive use of time-dependent orbitals as a single-particle basis set. For example the MCTDHB method, explained in Chap. 3 employs such orbitals. Also the time-dependent Wannier functions, introduced in Chap. 8 are of this type. We have always tried to stress the importance of time-dependent orbitals. In this Appendix we will show by direct comparison, why the use of time-dependent orbitals is crucial for the theoretical treatment of a many-boson system.

In dynamical problems time-dependent orbitals are generally complex functions, which makes them difficult to represent. We will therefore not consider a dynamical example, but rather one from statics, i.e. a stationary state. In both cases, dynamics and statics, time-dependent basis sets can adjust optimally to the interparticle interaction strength, the number of particles and the external potential, whereas time-independent basis sets cannot. The conclusions that we will draw from the statics example here therefore also apply to dynamical problems. In dynamical problems the use of time-dependent orbitals is actually even more important, since the numerical error can accumulate.

F.2 A Double-Well Ground State as an Example

We consider the ground state of $N = 1000$ interacting bosons in a 1D double-well potential of the form

$$V(x) = \frac{1}{2}x^2 + Ae^{-x^2/2\sigma^2}. \tag{F.1}$$

K. Sakmann, *Many-Body Schrödinger Dynamics of Bose–Einstein Condensates*, Springer Theses, DOI: 10.1007/978-3-642-22866-7,

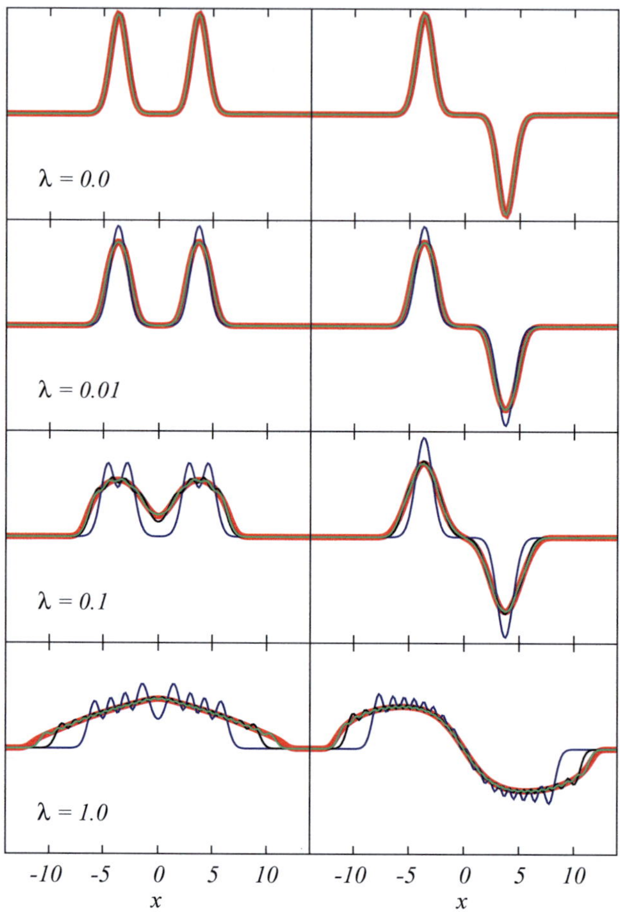

Fig. F.1 Expansion of natural orbitals in Wannier functions. Shown is the first (*left panels*) and the second (*right panels*) natural orbital of the ground state of $N = 1000$ bosons in a trap at different interaction strengths $\lambda = 0, 0.01, 0.1, 1.0$ (*from top to bottom*). MCTDHB (*thick red lines*) with $M = 2$ orbitals was used to obtain the natural orbitals. Also shown are the approximations ψ_k to the natural orbitals for different accuracies: $\epsilon = 0.1$ (*blue*), 0.01 (*black*), 0.001 (*green*). See text for details. Even at a medium accuracy of $\epsilon = 0.01$ a large numbers of bands κ is needed to represent the natural orbitals at any nonzero interaction strength: from top to bottom $\kappa = 1, 3, 7, 28$. The corresponding many-body basis set sizes are astronomically large, see Table F.1

As an interparticle interaction we choose $W(x - x') = \lambda_0 \delta(x - x')$. In Chap. 5 we also used the potential (F.1) and varied the barrier height A at constant interaction strength $\lambda_0 = 0.01$. Here we will vary the interaction strength at constant barrier height, $A = 22$ and $\sigma = 2$. For every λ_0 we compute the first two natural orbitals $\alpha_1^{(1)}(x)$ and $\alpha_2^{(1)}(x)$ using MCTDHB (with $M = 2$). We then expand these natural orbitals in time-independent Wannier functions as follows. First we obtain the

Table F.1 Many-body basis set size. Shown is the number of many-body basis functions if Wannier functions are used to expand the natural orbitals of $N = 1000$ bosons depicted in Fig. F.1. For all but the weakest interaction strength and the lowest accuracy the number of many-body basis functions is computationally unfeasible if Wannier functions are used as a single-particle basis set

Interaction strength λ_0	Accuracy ϵ	Number of bands κ	Basis set size
0.0	0.0	1	1001
0.01	0.1	1	1001
0.01	0.01	3	$\sim 8 \times 10^{12}$
0.01	0.001	3	$\sim 8 \times 10^{12}$
0.1	0.1	3	$\sim 8 \times 10^{12}$
0.1	0.01	7	$\sim 2 \times 10^{29}$
0.1	0.001	9	$\sim 3 \times 10^{36}$
1.0	0.1	16	$\sim 2 \times 10^{59}$
1.0	0.01	28	$\sim 3 \times 10^{92}$
1.0	0.001	33	$\sim 1 \times 10^{105}$

eigenstates of the noninteracting single-particle problem. These single-particle eigenstates form doublets which we call bands. We write α for the band index. The Wannier functions $w_j^{\nu}(x)$ with $j = L, R$ are then constructed as linear superpositions of the orbitals of each of the bands. There are two such Wannier functions per band. We now expand the previously obtained natural orbitals $\alpha_k^{(1)}$ in the Wannier basis and define the orbitals

$$\psi_k^{(\kappa)}(x) = \mathcal{N} \sum_{j=L,R} \sum_{\nu=1}^{\kappa} \langle w_j^{\nu} | \alpha_k^{(1)} \rangle w_j^{\nu}(x), \tag{F.2}$$

as approximations to the natural orbitals where $\mathcal{N}$ is a normalization constant. The approximation error ϵ using κ bands is then defined as the greater value of the norms of the differences between $\psi_k^{(\kappa)}$ and $\alpha_k^{(1)}$ for $k = 1, 2$

$$\epsilon = \max_k \left\{ \sqrt{\int dx \left| \psi_k^{(\kappa)}(x) - \alpha_k^{(1)}(x) \right|^2} \right\} \tag{F.3}$$

We will investigate how many bands are needed to approximate the natural orbitals $\alpha_k^{(1)}$ to a given accuracy.

Figure F.1 shows the natural orbitals $\alpha_k^{(1)}(x)$ together with their approximations $\psi_k^{(\kappa)}(x)$ at different interaction strengths for the accuracies $\epsilon = 0.1, 0.01, 0.001$. Even for the weakest nonzero interaction strength shown (second from above) at least $\kappa = 3$ bands are necessary to obtain a medium accuracy of $\epsilon = 0.01$. For stronger interaction $\kappa = 7$ and $\kappa = 28$ bands are needed then. The corresponding sizes of the many-body basis sets are $N + 2\kappa - 1N$ which are astronomically large. All numbers are collected in Table F.1.

The above discussion clearly demonstrates the necessity to use time-dependent basis sets for many-boson systems. Compared to time-dependent, optimized basis sets computations using time-independent basis sets are wasting computational resources at a very high rate with little chance to obtain converged results. We stress that the same conclusions can also be drawn from the analysis of dynamics problems, even at weak interaction strength.

Bibliography

1. A. I. Streltsov, O. E. Alon, L. S. Cederbaum, Phys. Rev. A **73**, 063626 (2006)
2. E. P. Gross, N. Cimento **20**, 454 (1961)
3. L. P. Pitaevskii, Zh. Eksp. Teor. Fiz. **40**, 646 (1961)
4. L. P. Pitaevskii, Sov. Phys. JETP **13**, 451 (1961)
5. C. J. Pethick H. Smith, *Bose-Einstein Condensation in Dilute Gases*, 2nd edn. (Cambridge University Press, Cambridge, 2008)
6. L. S. Cederbaum, A. I. Streltsov, Phys. Lett. A **318**, 564 (2003)
7. A. I. Streltsov, L. S. Cederbaum, N. Moiseyev, Phys. Rev. A **70**, 053607 (2004)
8. O. E. Alon, L. S. Cederbaum, Phys. Rev. Lett. **95**, 140402 (2005)
9. O. E. Alon, A. I. Streltsov, L. S. Cederbaum, Phys. Rev. Lett. **95**, 030405 (2005)
10. O. E. Alon, A. I. Streltsov, L. S. Cederbaum, Phys. Lett. A **347**, 88 (2005)

K. Sakmann, *Many-Body Schrödinger Dynamics of Bose–Einstein Condensates*,
Springer Theses, DOI: 10.1007/978-3-642-22866-7,

Biography
Publications, Conference Contributions, Software and Awards of Dr. Kaspar Sakmann

Publications

- *Accurate multi-boson long-time dynamics in triple-well periodic traps*, A. I. Streltsov, K. Sakmann, O. E. Alon, and L. S. Cederbaum, Phys. Rev. A **83**, 043604 (2011).
- *Optimal Time-Dependent Lattice Models for Nonequilibrium Dynamics*, K. Sakmann, A. I. Streltsov, O. E. Alon, and L. S. Cederbaum, New J. Phys. **13**, 043003 (2011).
- *Quantum dynamics of attractive versus repulsive bosonic Josephson junctions: Bose–Hubbard and full-Hamiltonian results*, K. Sakmann, A. I. Streltsov, O. E. Alon, and L. S. Cederbaum, Phys. Rev A **82**, 013620 (2010).
- *Exact Quantum Dynamics of a Bosonic Josephson Junction*, K. Sakmann, A. I. Streltsov, O. E. Alon, and L. S. Cederbaum, Phys. Rev. Lett. **103**, 220601 (2009).
- *Exact Quantum Dynamics of a Bosonic Josephson Junction*, K. Sakmann, A. I. Streltsov, O. E. Alon, and L. S. Cederbaum, Phys. Rev. Lett. **103**, 220601 (2009).
- *Reduced density matrices and coherence of trapped interacting bosons*, K. Sakmann, A. I. Streltsov, O. E. Alon, and L. S. Cederbaum, Phys. Rev. A **78**, 023615 (2008).
- *Rapidly driving a nanoparticle magnetic moment: Mean first-passage times and magnetic relaxation*, S. I. Denisov, K. Sakmann, P. Talkner and P. Hänggi, Phys. Rev. B **75**, 184432 (2007).
- *Mean first passage times for an ac-driven magnetic moment of a nanoparticle*, S. I. Denisov, K. Sakmann, P. Talkner and P. Hänggi, Europhys. Lett. **76**, 1001 (2006).
- *Exact ground state of finite Bose-Einstein condensates on a ring*, K. Sakmann, A. I. Streltsov, O. E. Alon, and L. S. Cederbaum, Phys. Rev. A **72**, 033613 (2005).
- *Continuous configuration interaction for condensates in a ring*, O. E. Alon, A. I. Streltsov, K. Sakmann, and L. S. Cederbaum, Europhys. Lett. **67**, 8, (2004).

K. Sakmann, *Many-Body Schrödinger Dynamics of Bose–Einstein Condensates*, Springer Theses, DOI: 10.1007/978-3-642-22866-7,

Contributions to Conferences

- *Exact Quantum Dynamics of a Bosonic Josephson Junction* DPG conference in Hannover 2010 – Q 10.8, Mo, Mar 8, 2010.
- *Exact Quantum Dynamics of a Bosonic Josephson Junction* Quantum Chemistry and Quantum Computational Physics in the Theory of Ultra-cold Gasses, Aug. 3-8, 2009, Telluride (Colorado), USA.
- *Exact Quantum Dynamics of a Bosonic Josephson Junction* LPHYS'09 18th INTERNATIONAL LASER PHYSICS WORKSHOP, July 13–17, 2009, Barcelona, Spain.
- *Reduced density matrices and coherence of trapped interacting bosons* Quantum Noise in Correlated Systems, Jan. 5-11, 2008, Weizmann Institute, Rehovot, Israel.

Software

- The MCTDHB Package, A. I. Streltsov, K. Sakmann, A. U. J. Lode, O. E. Alon, and L. S. Cederbaum, Version 2.1, Heidelberg, 2011, http://MCTDHB.uni-hd.de.
- OpenMCTDHB v2.1, K. Sakmann, A. U. J. Lode, A. I. Streltsov, O. E. Alon, L. S. Cederbaum (2011), http://OpenMCTDHB.uni-hd.de.

Awards

- Dr. Sophie-Bernthsen Award of the University of Heidelberg, 2008.

MIX
Papier aus verantwortungsvollen Quellen
Paper from responsible sources
FSC® C105338

If you have any concerns about our products, you can contact us on
ProductSafety@springernature.com

In case Publisher is established outside the EU, the EU authorized representative is:
Springer Nature Customer Service Center GmbH
Europaplatz 3, 69115 Heidelberg, Germany

Printed by Libri Plureos GmbH
in Hamburg, Germany